Oceans

by Ashlan Cousteau and Philippe Cousteau
with Joe Kraynak

for
dummies®

A Wiley Brand

Oceans For Dummies®

Published by: **John Wiley & Sons, Inc.,** 111 River Street, Hoboken, NJ 07030-5774, www.wiley.com

Copyright © 2021 by John Wiley & Sons, Inc., Hoboken, New Jersey

Published simultaneously in Canada

Contents at a Glance

Table of Contents

Introduction

The ocean covers about 71 percent of our planet's surface, contains about 97 percent of its water, is home to more than 90 percent of its living species, produces more than 60 percent of the oxygen on our planet, carries 90 percent of all cargo shipped between countries, produces enough protein to feed a billion people, contributes trillions of dollars to the global economy, regulates the climate, produces weather systems, provides us with all sorts of fun and interesting activities, inspires us, and so much more. Despite all this, we went ahead and totally dissed the ocean by calling our planet Earth . . . yeah, Earth as in land, ground, dirt. Wouldn't it have been much more appropriate to call it something like Planet Ocean, or maybe Oceanus, after the son of Uranus (Heaven) and Gaia (Earth)? We certainly think so.

But, nobody asked us, did they?

So, to help reconcile this slight in some small way, Philippe and I decided to write a book about the ocean and plaster its name all over the cover. We want everyone to know how truly amazing the ocean is and to share our fascination with it and its many crazy, astonishing, extraordinary, and sometimes creepy inhabitants. Our purpose in writing this book is to introduce generations of readers, including you, to the ocean from various perspectives — geophysics, biology, meteorology, ecology, economics, and more — in a fun and engaging way, so you can gain a broad understanding of what the ocean is all about without dozing off in the process.

As lifelong ocean explorers, we felt like this kind of primer on all things ocean, fun enough for the ocean novice but thorough enough for the ocean expert, was missing in the market. So, to plug that gap, we wrote this book and packed it chock full of photographs and illustrations because, well, words alone don't do the ocean justice, and because we're fortunate to have generous friends who are some of the most amazing photographers in the world and were willing to give us access to their archives.

Perhaps even more important is that we want to convey the vital role the ocean plays in humanity's survival; even if you live in the middle of the desert, the ocean still affects you and your quality of life. Unfortunately, the ocean is taking the brunt of humanity's ecological abuses, which spells trouble for all of us. The good news is that each and every one of us has the power to save it. By working together to stop and perhaps even reverse the damage we've caused while optimizing our use of the ocean's vast resources, we can return the ocean to abundance. Our intent isn't to be Debbie or Danny Downer. Our goal is to have fun, engage you, and maybe even recruit you to join the cause of loving and protecting the sea.

So, tighten up your swimsuit, strap on your scuba gear, and get ready to dive into what we feel is the most fun and fascinating of topics, the ocean.

About This Book

Welcome to *Oceans For Dummies*. In this book, we serve as your personal tour guides to Earth's watery wonderland. We lead you on a journey from shore to open ocean and from surface to seafloor, examining the ocean from multiple perspectives — its history, biology, ecology, economy, and much more. Along the way, we hope to expand your appreciation and general understanding of the ocean and its inhabitants. We also hope to increase your awareness of the opportunity to interact with the ocean in mutually beneficial ways, so we can all enjoy its vast resources while preserving it for future generations.

To make the content more accessible, we divided it into the following six parts:

>> Part 1, "Getting Started with Your Ocean Voyage," touches on key topics covered throughout the book, highlights the many ways the ocean enhances our lives, and traces the origin of the ocean and the evolution of marine life (yeah, all that in three short chapters).

>> Part 2, "Finding Your Way Around," divides the ocean into zones and *ecosystems* (communities of interdependent marine organisms) and takes a deep dive to the seafloor to scope out what's on, in, and below it — you may be surprised.

>> Part 3, "Sampling the Vast Diversity of Sea Life," consumes a vast majority of the book because, frankly, it's the topic we find most fun and fascinating. Here, you meet shrimp that can smash open aquarium tanks with their bare knuckles; the brainy octopus, which can twist the lid off a jar using its suction-cupped grip (very handy in the kitchen); the largest creature on the planet, *ever,* which feeds on the smallest of prey; and numerous other creepy, freaky, and awesome characters.

>> Part 4, "Grasping Basic Ocean Physics," looks at the ocean as a part of an integrated system comprised of water, land, atmosphere, the sun, the moon, and certain physical forces that make Earth what it is. In this context, we explain how water, heat, and nutrients are circulated around the ocean and around the globe and how the ocean helps to regulate climate and drive weather.

>> Part 5, "Understanding the Human-Ocean Connection," focuses on the many ways the ocean benefits our lives, how the exploitation of its resources are governed, and how we humans need to transition our relationship with the ocean from one of dependence to interdependence to fully enjoy its benefits while preserving its health. In this part, we also dip into the topic of ocean exploration — showcasing the innovations and technologies that have driven its development over the last couple hundred years.

>> Part 6, "The Part of Tens," features three "top ten lists" — the ten deadliest sea creatures, ten ocean myths we bust wide open, and ten easy ways you can help to preserve the ocean for future generations.

In short, this book serves as your guide to almost everything about the ocean in a fun and accessible way. And while some of these subjects deserve an entire book for themselves (hello, nudibranchs!), we worked hard to find the balance between breadth and depth.

Foolish Assumptions

All assumptions are foolish, and we're always reluctant to make them, but to keep this book focused on the right audience and ensure that it delivers the information and insight you need to grasp a topic as broad as the ocean, we had to make the following foolish assumptions about you:

>> You're curious about the ocean and eager to discover more about it.

>> You're mostly interested in marine life, such as turtles, sharks, and dolphins and ecosystems, such as coral reefs and kelp forests.

>> You're concerned about the health of the ocean, and you want to find out more about what can be done, what's being done, and what you can do to help the ocean thrive.

Other than those three foolish assumptions, we can honestly say that we can't assume much more about you. For all we know, you could be a precocious 4-year old or a spirited octogenarian, you may be a student, a white collar or blue collar worker, a housewife or househusband, a doctor, a lawyer, a plumber or a construction worker. You could be living on an island, a coast, deep in the Amazon jungle, or on a farm in the middle of Iowa. Regardless of your demographic, we applaud your interest in the ocean and your eagerness to expand your knowledge and understanding of Earth's most precious blue gem.

Icons Used in This Book

Throughout this book, icons in the margins highlight certain types of valuable information that call out for your attention. Here are the icons you'll encounter and a brief description of each.

REMEMBER

We want you to remember everything you read in this book, but if you can't quite do that, then remember the important points flagged with this icon.

The ocean is full of surprises both entertaining and mind blowing, and we don't want you to miss any of them, so we use this Fun Fact icon to flag every single one.

We don't advise you to do anything very dangerous in this book so you won't bump into many of these warning icons. In fact, we even try to alleviate any fears you may have about potentially dangerous marine creatures like big bad sharks (ooohhh), jellyfish, and sea snakes. Most of what you read in the news and see in the movies is way overblown. However, we did have a couple opportunities to use this icon, so watch out for it.

Beyond the Book

In addition to the ocean of information and insight we provide in this book, you have access to even more help and information online at Dummies.com, including a Cheat Sheet that serves as a quick reference guide to this book. We also posted a list of our ten favorite sea creatures — a bonus item that truly goes "beyond the book." To access the Cheat Sheet, go to www.dummies.com and search for "Oceans For Dummies Cheat Sheet."

Where to Go from Here

You're certainly welcome to read this book from cover to cover, but we wrote it in a way that facilitates skipping around. For a quick cruise around the ocean, turn to Chapter 1. To find out how the ocean formed and how marine life evolved (how the ocean as we know it came to be), head to Chapter 3. To scope out different areas of the ocean, head to Part 2, where we divide the ocean into zones and ecosystems and check out the seafloor.

All chapters in Part 3 are required reading, but feel free to just flip through them at your leisure to look at all the pretty pictures. Words just can't capture the beauty of the ocean and the creatures that call it home as well as photos can. Chapter 21 is also required reading. One of the primary reasons we wrote this book is to recruit you to join our mission to save the ocean — one of the world's most precious resources. In Chapter 24, we provide a menu of easy ways you can help.

Other than that, you're pretty much on your own to explore at your own pace and as your curiosity leads you. Enjoy!

1

Getting Started with Your Ocean Voyage

Take a quick primer on ocean fundamentals — from dividing the ocean into oceans (plural) and getting up to speed on the water cycle to engaging in a quick meet-and-greet with the ocean's inhabitants.

Take inventory of the various ways the ocean contributes to our health and happiness, not to mention our very existence.

Discover how the ocean and oceans formed and how life on Earth may have begun.

Trace the evolution of marine life from the Paleozoic to the Cenozoic era and everything in between.

Look ahead to find out what the future of the ocean might look like.

Chapter **1**

Brushing Up on Ocean Fundamentals

Before you dive into any large or complex topic, you're wise to step back and look at the big picture. A general understanding provides a framework on which to hang the details. In this chapter, we provide that framework, establishing a context for understanding the many facets of the ocean and how they all fit together.

We start by introducing you to the ocean and breaking it down into its five "oceans." Then, we get into a few topics in the realm of *physical oceanography* — the water cycle, the shapes of the ocean basins, meteorology, and other properties and processes that explain what makes the ocean tick from a physical standpoint. We then introduce you to the various groups of life-forms that populate the ocean — plants, animals, and beings that fit in neither (or both) categories. Finally, we wrap things up with a discussion of the ocean's current state and the human-ocean relationship — the benefits we gain from the ocean and our responsibilities as environmental stewards in protecting and preserving it.

Get ready for a wild ride. In this chapter, we cover a lot of ground, not to mention all that water!

Taking a Nickel Tour of the Ocean(s)

The ocean is big. How big? Well, it covers about 360 million square kilometers (140 million square miles), which is just a smidgen more than 70 percent of Earth's surface. Volume-wise it contains approximately 1.3 billion cubic kilometers (321 million cubic miles) of water — that translates to about 352 quintillion gallons, which accounts for about 97 percent of Earth's water. In terms of living space, the ocean comprises about 99 percent of the *biosphere* — all land, water, and atmosphere where life on Earth exists.

Because it's so big, people have developed all sorts of ways to slice and dice it to better understand and describe the different areas that make it up.

Dividing the ocean into oceans . . . or not

Earth has only one ocean, which is why we will refer to it as *the ocean* (singular) throughout this book. Geographically, it's divided into four or five oceans, depending on who's doing the dividing. Prior to the year 2000, text books recognized four oceans: the Atlantic, Pacific, Indian, and Arctic. Sometime around the year 2000, the International Hydrographic Organization designated a fifth ocean the Southern Ocean — a band that wraps around the world from the coast of Antarctica to 60 degrees south latitude (see Figure 1-1). Here's a brief description of each of the five oceans, in order of size, because, well, size matters.

- » **Pacific Ocean:** The largest of the five "oceans," the Pacific stretches from the Arctic to the Southern Ocean and from east of Asia and Australia to the Americas. It covers more area than all the land on Earth combined and is more than double the surface area of the Atlantic Ocean. It also wins the deepest point in the ocean contest with the Mariana Trench, which is nearly 11 kilometers (about 7 miles) deep.

- » **Atlantic Ocean:** This next largest ocean lies between the Americas and the continents of Europe and Africa. It's home to the Bermuda Triangle, the Sargasso Sea, the Gulf Stream, and the hurricanes that rattle the Caribbean Islands and the southern and eastern coasts of the U.S. The North Atlantic is *by far* the most thoroughly explored, best understood, and most heavily fished of the five "oceans."

- » **Indian Ocean:** Nestled between Africa (to the west) and Australia (to the east) and between Asia (to the north) and the Southern Ocean (below it), the Indian Ocean ranks third in surface area but first in warmth.

>> **Southern Ocean:** The Southern Ocean is relatively small, but its average depth is greater than the average depth of any of the other four oceans — four to five kilometers (2.5 to 3 miles) deep! It's best known for its strong, sustained easterly winds, its huge waves (due to the strong, sustained winds), and its frigid environment; during its winter, nearly the entire surface of the Southern Ocean is frozen. It's also home to the world's largest ocean current — the Antarctic Circumpolar Current — and it is chock full of nutrients.

>> **Arctic Ocean:** Surrounding the North Pole and bordering the northern edges of North America, Asia, and Europe is the Arctic Ocean. Most of it is located within the Arctic Circle, from the North Pole down to about 70 degrees northern latitude. It's the smallest and shallowest of the five "oceans," and for most of the year, most of its surface area consists of ice 1 to 10 meters (3 to 33 feet) thick. The Arctic Ocean is best known for its wildlife (including polar bears, whales, and seals) and for its natural resources (primarily oil).

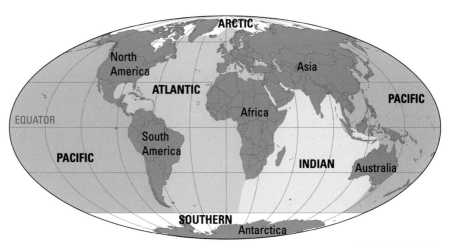

FIGURE 1-1:
The ocean's five "oceans."

©John Wiley & Sons, Inc.

Recognizing the ocean zones

Oceanographers have divided the ocean into zones to better understand and describe the physical characteristics of the ocean, the *ecosystems* (biological communities) in each zone, and the inhabitants of those ecosystems. Zoning can be simple, such as dividing the ocean into two zones — photic and aphotic:

>> **Photic (light):** The top 200-meter (650-foot) layer of the ocean through which enough light penetrates enabling photosynthesis to occur. (*Photosynthesis* is the process of using the sun's energy to produce food from carbon dioxide and water.)

>> **Aphotic (dark):** The part of the ocean from 200 meters down to the bottom, where it's totally dark.

Another simple zoning system involves dividing the ocean into pelagic and benthic layers:

>> **Pelagic (top):** The water above the ocean floor.

>> **Benthic (bottom):** The seafloor and the thick layer of sediments below the seafloor.

In Chapter 4, we cover two more-detailed approaches to zoning the ocean — one that divides it into five horizontal layers (like layers of a cake) based on depth, and another that divides it into three vertical zones from coast to open ocean.

Dropping in on the different ecosystems

An *ecosystem* is a biological community of organisms interacting with their physical environment as a whole. Think of it as a mostly self-contained, self-reliant neighborhood with a diverse population. Land-based ecosystems include grasslands, deserts, rainforests, and wetlands. Common marine ecosystems include coral reefs, estuaries (where fresh water and salt water mix), kelp (seaweed) beds and forests, mudflats, rocky shores, sandy shores, seagrass meadows, and more. Lesser known ecosystems develop near the bottom of the deep sea and include communities that form around *hydrothermal vents* (which spew hot, mineral-rich water that some bacteria feed on), *whale falls* (literally, dead whales that sink to the bottom), and *cold seeps* (where methane gas is released that some bacteria and archaea feed on).

What's so fascinating about ecosystems is that the community of residents that live within them evolved together, adapting to the unique conditions of a particular place as well as each other. In Chapter 5, we explore numerous marine ecosystems and introduce you to the plants, animals, and other organisms typically found in each.

Going with the Flow: The Physical Properties of the Ocean

Even without all the wonderful living organisms that call it home, the ocean is amazing. In fact, it is very much like a living thing itself; it breathes, it moves, it's constantly changing, and it interacts with everything around it — land, water, and air. It plays a huge role in controlling Earth's climate and making the weather, and it distributes heat and nutrients around the globe, making the entire world more habitable for every living thing.

In this section, we get physical by focusing on the salt in seawater, the various processes that maintain a steady flow of water around the world, and the interactions among land, sea, and air.

Getting up to speed on the water cycle

While plants and animals have life cycles, water has a life cycle of its own. Following the rule of "what goes up must come down," the life cycle of water, commonly referred to as the *hydrologic cycle,* describes the way water travels around the globe from ocean to air to land and back again to the ocean (see Figure 1-2).

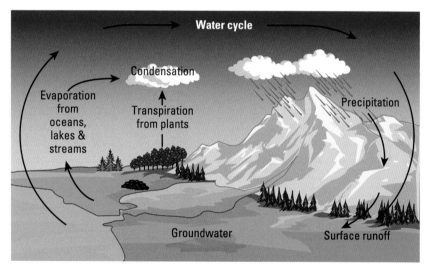

FIGURE 1-2:
The hydrologic (water) cycle.

Water is a wonderous element which can take on three different states of being. It can exist in a solid state (ice), a liquid form (water), or a gaseous state (steam), in which it evaporates and becomes vapor (humidity). When humid air is cooled, the water forms droplets and falls back to the earth as *precipitation* — in either liquid form (as rain) or in solid form (as snow or hail).

Knowing what makes seawater salty

Why are most lakes, rivers, ponds, and streams all freshwater, whereas the ocean is salty? Mostly because of the water cycle. Most of the salt in the ocean comes from freshwater rivers and runoff from land. As the water moves over the land and rocks and through the earth, it picks up minute traces of salt and other minerals, which it then carries to and deposits in the ocean.

When water evaporates from the ocean, the salt remains, while the water vapor either precipitates over the ocean or is carried over and deposited as precipitation on land, where it washes more minerals into the ocean. This process "quickly" (that is, on a geological time scale) increased the concentration of salt in seawater to the level at which it is today — on average about 3.5 percent.

Investigating variations in salt concentrations

Because water evaporates faster from some parts of the ocean than others and rainfall and river discharge to the ocean isn't uniform around the globe, certain parts of the ocean are saltier than others. For example, salinity in the Red Sea ranges from 36 ppt (parts per thousand) to 41 ppt (about 3.6 to 4.1 percent). In the Mediterranean Sea, surface waters average 38 ppt but can approach 40 ppt in the extreme western parts. Saltwater taffy, anyone?

Salinity also fluctuates in certain areas of the ocean. For example, in polar regions, the salinity of water near the surface increases as ice forms, leaving much of the salt behind. As the ice melts, salinity decreases due to the influx of fresh water. Likewise, in coastal areas, salinity is lower in areas where freshwater from rivers and streams enters the ocean. When the salinity dips below about 30 ppt, the water is often called *brackish*.

Realizing that sea water is more than just salty water

The recipe for cookin' up a batch of seawater might seem quite simple: Start with 1 gallon (about 4 liters) of water, add about 4 tablespoons (60 grams) salt, and shake till dissolved. However, seawater is more complex than mere salt and water — it contains a very large number of chemical elements and compounds, including dissolved organic materials, particulates, and dissolved gasses, including oxygen. (Marine mammals can't breathe the gasses dissolved in seawater, but apparently the crew of the movie *The Abyss* can.)

Don't worry, we're not going to deliver a lecture on the chemical composition of seawater, but knowing that seawater is more than just salty water is important for when we get into discussions about certain issues related to preserving the ocean's health, such as climate change and ocean acidification (see Chapter 21).

Checking out what's at the bottom of the ocean (and below)

At the bottom of the ocean is a landscape that's as varied as what you see above the ocean, complete with plains, ridges (mountain ranges), seamounts (mountains), hills, canyons, trenches (valleys), cliffs, volcanoes, hydrothermal vents (underwater geysers), and even rivers and massive "waterfalls" of sand and sediment that you do indeed want to go chasing.

The shape of the ocean floor (as well as all land) is largely determined by a process called *plate tectonics* — the bumping, grinding, and separating of *tectonic plates* (massive sections of rock that form Earth's crust). These plates ride atop a semi-liquid layer of molten rock (imagine heating a rock till it melts — that's HOT). Convection currents in the molten rock move the plates at three to five centimeters (one to two inches) per year. (See Chapter 6 for more about plate tectonics.)

Plate tectonics are also responsible for creating most tsunamis. Whenever land shifts at the bottom of the ocean or an underwater volcano erupts, the resulting displacement of water can send a wave thousands of miles across the ocean in a matter of only a few hours. (See Chapter 16 for more about waves and tsunamis.)

Riding waves, tides, and currents

Ocean water is constantly in motion thanks to a variety of forces, including the rotation of Earth; the gravitational fields of the sun, Earth, and moon; wind; and variations in the temperature and salinity of ocean water.

On a cosmic scale, the interaction of the Earth's, moon's, and sun's gravitational forces cause *tides* that move water toward and away from coastlines once or, more often, twice daily.

At the ocean's surface, wind, along with other forces (including Earth's rotation and variations in water temperature and salinity) drive the formation of large rotating waters called *gyres* that generally spin clockwise in the Northern Hemisphere and counterclockwise in the Southern Hemisphere. (And no, this has no effect on the direction water swirls down your bathtub drain.)

Below the surface, differences in water density drive currents. Near the poles, when water freezes at the surface, it leaves behind its salt. This colder, saltier water near the surface is denser than the warmer, fresher water below it, so it dives down, hits the seafloor, and heads toward the equator. As this cold water moves down and away from the poles, warmer surface water rushes in toward the poles to replace it, resulting in a continuous current that transports water, heat, and nutrients around the globe. This process is called *thermohaline circulation*, and the resulting system of currents is called the *global ocean conveyor belt*. It is one of the many reasons the poles are so important. They literally drive our entire ocean system, supporting marine life (and seafood) all around the globe.

See Chapter 16 for more about waves, tides, currents, and gyres and how water, heat, and nutrients are circulated around the world via ocean currents.

Recognizing the ocean's role in climate control and weather

The ocean plays a critical role in keeping our planet at a steady temperature as it transports heat (and energy) around the globe. Of course, it's not that simple, and the ocean doesn't do this all by itself; it works together with the atmosphere and land to create our climate and influence the ever-changing weather patterns:

>> **Climate:** The prevailing weather over a long period of time (usually more than 30 years), such as tropical (warm and wet), desert (hot dry), polar (cold and dry), and temperate (neither extremely hot nor extremely cold).

>> **Weather:** Atmospheric conditions over a short period of time in respect to temperature, sunshine, storms, wind, and precipitation (rain, snow, sleet, hail).

The impact of the ocean on weather is most dramatic when the ocean releases some of its energy in the form of a *tropical cyclone* (a hurricane, cyclone, or typhoon). See Chapter 17 for more about how the ocean influences climate and weather.

Meeting the Ocean's Inhabitants

The ocean is teeming with life, from coastline to open ocean and from surface to seafloor. These living beings can be broken down into six groups (technically known as *kingdoms*) — Plants, Animals, Protists, Fungi, Archaebacteria, and Eubacteria. In this book, we break them into three groups — microbes (generally too small to see with the naked eye), plants (and other organisms that require sunlight for energy and growth), and animals. Each of these groups can be broken down further; for example, the animal kingdom includes simple invertebrates (such as sponges and jellies), mollusks (such as snails and clams), crustaceans (such as crabs and lobsters), fish, reptiles, birds, and mammals.

In this section, we cover the bare bones of *taxonomy* — the classification system used to assign organisms to specific groups — and introduce you to the groups of organisms we cover in Part 3.

Recognizing strength in numbers: Marine microorganisms

Microorganisms are (mostly) life-forms too small to see with the naked eye. Many of these are smaller than the cells that make up our bodies, but what they lack in size, they more than make up for in numbers. The ocean is home to an estimated 44 *octillion* microbes — more than all the stars in the known universe. Perhaps more amazing is that microbes comprise somewhere between 90 and 98 percent of the marine biomass (the total mass of all marine life)! These microorganisms can be broken down into four groups:

>> **Viruses** are infectious agents that invade cells and use them to replicate. They're not even classified as living beings (*burn!*).

>> **Bacteria** are single-celled organisms that have a cell wall but no *nucleus* (a container for most of the cell's genetic material) or *organelles* (specialized structures in a cell that perform various functions).

>> **Protists** are single-celled organisms equipped with a nucleus and organelles. They're not animals, plants, or fungi, but some are similar to plants in that they perform photosynthesis, some are similar to animals in that they move around and eat stuff, and some are more of a cross between the two — moving around, eating stuff, *and* performing photosynthesis.

>> **Fungi** are single-celled or multicelled organisms that feed on organic matter. They include yeasts, molds, mushrooms, and toadstools.

REMEMBER

Microorganisms are a vital component of oceans. They serve as *producers*, making food for consumers at the bottom of food chains, and they serve as *decomposers*, breaking down animal waste into chemicals that can be reused. See Chapter 7 for more about microbes and the important roles they play.

Going green with marine plants and plant-like organisms

Oceans aren't exactly known for their floral arrangements, but they do have a few recognizable plants and plenty of other important photosynthetic microorganisms that provide food, oxygen, and habitats for other marine organisms. Here's a list of some of the more notable marine plants and photosynthetic organisms covered in Chapter 8:

>> **Phytoplankton** are single-celled photosynthetic organisms sometimes described as "floating plants." Most are single-cell algae (micro-algae), but phytoplankton also include a type of photosynthetic bacteria called *cyanobacteria*.

>> **Macro-algae** are several species of macroscopic, multicellular marine algae commonly referred to as "seaweed." They look like plants and can be very large, but they have no vascular system for distributing nutrients. All the cells that make up the seaweed absorb liquids and nutrients from the surrounding water.

>> **Seagrass** is a plant, complete with a vascular system, leaves, roots, and *rhizomes* (just like lawn grass); they're pollinated under water, and they produce seeds.

>> **Mangroves** are highly salt-tolerant trees and shrubs that grow along shorelines in tropical locations. In addition to providing food, shelter, and breeding areas for other marine organisms, mangroves play an important role in protecting and even building land.

>> **Zooxanthellae:** Pronounced zo-uh-zan-*thel*-ee, these are single-celled photosynthetic organisms that live inside the cells of many marine animals, including coral polyps and some jellyfish, nudibranchs (sea slugs), and sponges, providing them with food in exchange for a place to live.

Grouping the ocean's animals

Although microbes and plants play a vital role in keeping the ocean clean and fed, they don't draw the crowds. The stars of the show are the animals, and the cast of marine creatures is truly incredible.

Unfortunately, this book can't possibly cover all the amazing creatures that inhabit the ocean, so what we've done is break them down into taxonomic groups and subgroups, describe the common traits of each group, and then present one or more representatives of each group. We cover the following groups, progressing from least to most complex, and point out where to find them in this book:

>> **Simple invertebrates** are basic animals that have no backbone, including sponges, jellyfish, anemones, starfish, sea urchins, sand dollars, and a few different types of worms (yes, worms) that prefer the ocean over your lawn or garden. Wiggle over to Chapter 9 for more in simple invertebrates.

>> **Mollusks** (covered in Chapter 10) are soft-bodied invertebrates, most of which have recognizable shells but some of which don't. You may know them better as snails, slugs, clams, oysters, mussels, octopus, and squid. And some of these are colossal.

>> **Crustaceans** (Chapter 11) are more advanced invertebrates with hard external skeletons, such as crabs, lobsters, shrimp, and krill. Just think "crusty," like a good baguette: hard on the outside, soft on the inside.

- » **Fish (bony and not)** comprise the first group of vertebrates (animals with a backbone or something like it). You usually know a fish when you see one — most have a head and tail, fins, gills, and scales. We break them down into two groups: *cartilaginous* (the real softies, as in soft-boned) such as sharks and rays, and *bony* (hard boned, that is) which includes just about everyone else, such as tuna, salmon, and cod. Swim over to Chapter 12 for more details.

- » **Reptiles** (chillin' in Chapter 13) are scaly, cold-blooded, air-breathing vertebrates that lay soft eggs on land or give birth to live young. The ocean is home to only a handful of reptiles, including sea turtles, marine iguanas, saltwater crocodiles, and a few species of sea snakes (the latter of which you really don't want on your airplane).

- » **Birds** are warm-blooded, air-breathing vertebrates with two legs, two wings, feathers, and a beak. Most fly, though some, such as the penguin, don't. Most marine birds have special adaptations, such as the ability to secrete salt, oily wings to keep them from getting waterlogged, and webbed feet. Some have solid bones that enable them to dive more easily. Seabirds that dive-bomb their prey are even equipped with internal airbags to cushion their crash landings. In Chapter 14, we divide marine birds into two groups — seabirds, which spend most of their time on or flying over the ocean, and shorebirds, which spend most of their time on or near land or wading in the shallows of estuaries or marshes.

- » **Mammals** (covered in Chapter 15) are warm-blooded vertebrates that have at least some hair or fur and must surface to breathe air. Females have mammary glands and give birth to live young. Marine mammals include everyone's favorites — whales, dolphins, walruses, seals, sea otters, sea lions, manatees, dugongs, and polar bears.

Exploring the Complex and Evolving Human-Ocean Relationship

Every relationship requires some give and take, and this is certainly true regarding the relationship between humans and the sea. For millions of years, humans have been taking from the ocean without giving much in return. We've used it for food, transportation, vacation, recreation, and exploration. In the past 100 years or so, we've come to rely on it as a source for energy, minerals, and medications. And the ocean has always been a source of mystery, inspiring scientists, artists, and writers. (See Chapter 19 for more about the many ways the ocean makes a positive impact on our lives — economically and in other ways.)

Tragically, the ocean and its inhabitants are suffering, and it's time we give back to the ocean which has given us so much. We humans have caused most of the ocean's problems — ocean warming and acidification from greenhouse gas emissions, a steep decline in fish populations, devastating pollution like plastic and runoff, harmful coastal development resulting in the destruction of marine habitats, mining and drilling, and the introduction of invasive species.

Despite all these negatives, we have high hopes for a brighter future. The ocean is resilient and can help us solve many of these problems, but we all need to work together to reduce and reverse the damage. Here are some of the steps we must take to preserve this precious and irreplaceable resource:

>> **Achieving net-zero greenhouse gas emissions:** Human-caused greenhouse gas emissions include dioxide, methane, ozone, nitrous oxide, and chlorofluorocarbons. Achieving net-zero greenhouse gas emissions by 2050, at a minimum, is critical to slowing and stopping human-caused climate change, ocean acidification, and warming ocean temperatures.

>> **Reduce pollution:** Pollution in the air, water, and even on land ends up in the ocean. Sources of pollution include plastics (which never really biodegrade), sewage, industrial chemicals, agricultural fertilizers and pesticides, land runoff (especially from streets and parking lots), oil spills, ocean mining, littering, and the use of certain sunscreens.

>> **Fish sustainably:** Overfishing combined with destructive fishing methods are devastating marine wildlife populations risk. International legislation and enforcement are crucial for ensuring a sufficient supply of seafood for ourselves as well as marine ecosystems.

>> **Create marine protected areas (MPAs):** MPAs are areas where certain consumptive or destructive activities are prohibited for the purpose of protecting ecosystems, sustaining fisheries production, or preserving cultural resources.

>> **Engage youth:** The best way to make lasting change is from the ground up. It starts by educating the youth. They're best equipped to change their own behaviors and influence the minds, hearts, and actions of parents, teachers, government representatives, community members, and business leaders.

For more about threats to the ocean and what we can do to preserve it for future generations, see Chapter 21.

Pretty cool, huh? The ocean is an amazing place, and we hope that reading this book makes you fall in love with the ocean for the first time . . . or all over again; that you learn, laugh, maybe even shed a tear; and, most importantly, that you're inspired, knowing *YOU* have the power to help our mighty ocean.

Chapter **2**

Appreciating the Ocean's Many Gifts

The ocean covers 71 percent of the planet, contains 97 percent of the planet's water, is home to 50 to 80 percent of all life on Earth (94 percent of life on Earth is aquatic), and supplies at least 50 percent of the world's oxygen, (scientists estimate it's anywhere from 50 to 80 percent, according to NOAA). It serves up about 200 billion pounds of seafood each year and supplies an important source of protein for billions of people around the world. It plays a big role in transportation, both travel and shipping, and a huge role in regulating climate and weather and making Earth habitable. It is drilled for crude oil, mined for minerals, tapped for development of new medications, and desalinated to provide drinking water. It is home to some of the most beautiful and amazing creatures on the planet and is an unfathomable source of wonder. We literally can't live without it.

The ocean is truly a gift that keeps on giving. Every planet wants one, other planets and moons across the universe probably have them, but Earth is the only planet we know of that has such a healthy, vibrant ocean teeming with life.

In this chapter, we look at the many ways the ocean makes Earth habitable and such a wonderful place to live, not to mention all the valuable resources it provides to us. Think of this chapter as a celebration of Ocean Appreciation Day.

REMEMBER

Actually, there is no Ocean Appreciation Day, but there is a World Oceans Day, when people around the world celebrate the ocean, commit to restoring and protecting it, and join in a number of conservation activities and events around the world. We encourage you to celebrate this day annually on June 8. (Visit worldoceansday.org for details.)

Supplying Over Half of the World's Oxygen

You've probably heard the Amazon described as "the lungs of our planet." However, while this incredible ecosystem is very important, rainforests supply only about 28 percent of the oxygen on our planet. That's nothing to sneeze at, unless, of course, you're allergic to rainforests. So where does the other roughly 70 percent come from? Well, over half of all oxygen produced on Earth comes from the sea (by some estimates it is as high as 80 percent!). While most of it is consumed by marine organisms, a small fraction of it escapes and over geologic time has given rise to an atmosphere that is about 20 percent oxygen, which has made life as we know it possible. Thank you, ocean. And this is mostly thanks to microscopic plantlike organisms floating in the sea, the tiny but mighty phytoplankton. (See Chapters 7 and 8 for more about phytoplankton and other ocean plants.)

Powered by the sun, these microscopic wonders turn carbon dioxide and water into sugar and oxygen through the process of photosynthesis, as do plants that grow on land (and in your flower pots), but it performs this magic on a much larger scale. According to some estimates, phytoplankton account for only one percent of all the biomass on our planet. Now, that may be a tiny slice of the pie, but they are able to conduct nearly as much photosynthesis as all the land plants that account for a much larger fraction of that pie. Without phytoplankton creating oxygen in the ocean, life on Earth, in its current form, could not exist.

Playing a Key Role in Regulating Climate and Weather

The ocean is like a massive version of the heating, ventilation, and air-conditioning (HVAC) units used to heat and cool most homes, but instead of circulating hot and cold air, it captures heat from the sun and distributes it through the circulation of water. Imagine Earth's frozen poles as air conditioners, the equator as its heater, and the ocean as the blower and ductwork, distributing the heat and cold all around the world.

Ocean currents transport warm water from the equator to the poles and cold water from the poles to the tropics. These currents act like a gigantic conveyor belt, moving warm and cool water to areas of contrasting temperature, thereby keeping the entire planet at a fairly comfortable 58.3 degrees Fahrenheit (on average). Sure, that's sweater weather for people and might make a polar bear break out in a sweat, but as an average temperature it's perfect.

Without ocean currents, regional temperatures would be more extreme — super hot at the equator and far more frigid than it already is at the poles — and much less of Earth's land would be habitable. See Chapter 16 for more about how the ocean circulates water and Chapter 17 for how it influences climate and weather.

DIFFERENTIATING CLIMATE FROM WEATHER

Nobody has trouble with the concept of weather because it impacts us every day. We check weather forecasts daily to plan our outdoor activities, to find out how to dress ourselves and our kids, to decide whether or not to carry an umbrella, and more. On the other hand, many people struggle with the concept of climate. For example, some people wonder how climate change (or global warming, as it is sometimes called) can be real when they experience a rogue snowstorm in the spring.

Well, the difference between weather and climate is that weather is affected by short-term changes in the atmosphere, which can still cause a cold spell when there shouldn't be one, whereas climate describes the AVERAGE weather in an area over a LONG PERIOD of time. For example, weather in a desert may be rainy or sunny on a particular day, but the climate is dry. Likewise, most of South Florida has a tropical climate (hot and humid), but the weather on some days may be cool and dry relative to Florida's seasonal climate.

So let's be very clear: Climate change (global warming) is real. According to an ongoing temperature analysis conducted by scientists at NASA's Goddard Institute for Space Studies (GISS), the average global temperature on Earth has increased by more than 1 degree Celsius (1.8 degrees Fahrenheit) since 1880. That may not seem like much, but it is. Head over to Chapter 17 for more about climate change (global warming) and how it's bad news for the ocean and for us.

The ocean also plays a key role in the *water cycle* — the continuous process by which water is carried around the world through evaporation, condensation, *precipitation* (rain and snow), and *transpiration* (the movement of water through plants). Even though the ocean is salt water, water that evaporates from the ocean's surface is fresh water, and much of it falls to the ground as rain or snow, bringing us essential water for drinking and for growing food.

Nearly all precipitation that falls on land originates in the ocean.

The water that evaporates combined with heat from the ocean's surface is responsible for the powerful storms that unleash their energy over land, often damaging coastlines and destroying property, but this is all part of the ocean's role in regulating the weather and contributing to the water cycle. By the way, the ocean's surface temperature can also impact the severity of storms, making them weaker or stronger.

Producing Protein for Billions of People

Humans and our ancestors have been eating seafood for a looooooooooong time. Yep, in a cave called Figueira Brava, located outside Lisbon (Portugal), remains of harvested mussels date back to a Neanderthal dinner some 80,000 to 160,000 years ago. Evidence also shows that Homo sapiens harvested shellfish at South Africa's Pinnacle Point between 164,000 and 120,000 years ago. So our love of seafood and its important nutrients goes way back.

According to the World Wildlife Fund, three billion people around the world today depend on wild-caught or farmed seafood as their primary source of protein. Now that's a whole lot of sushi (and other seafood), and it doesn't count consumption by the other five billion or so people on the planet. That makes seafood the largest traded food commodity on Earth. But this incredible ocean bounty is being pulled out at unsustainable rates; 90 percent of all fish stocks today are either overfished or fished to capacity (see Chapters 19 and 21 for details).

To feed the nine or close to ten billion people estimated to populate the planet by 2050, we need to make sure our fisheries are healthy and brimming with fish. According to one estimate (oceana.org/feedtheworld), with the right management and restoration, we could increase global fish stocks by 15 percent, enough fish to feed one billion people a seafood meal every single day.

Save the fish! One way you can contribute to a sustainable ocean is by switching supplements. Currently, fish and krill are a main source of omega 3s in most supplements. But fish and krill get their omega 3s from algae, so all you need to do is cut out the middleman by switching from fish oil to algae oil. Algae pills provide the same omega 3 benefits, but they're vegetarian, tasteless (no nasty fish burps after taking them), and ocean-friendly. Why make the switch? Well, the fish oil/krill oil industry is wreaking havoc on the marine food web. See Chapter 21 to find out more about what can be done to restore marine life populations and diversity.

Contributing Trillions to the Global Economy

Sticking a price tag on the ocean is a real challenge. How would you go about measuring its economic value? Would you use a metric like the one used for countries — gross domestic product (GDP)? And if the value of the ocean could be quantified, would it be considered rich compared to the countries with the biggest economies?

Well, in 2015, the World Wildlife Fund, the Global Change Institute (at the University of Queensland), and the Boston Consulting Group set out to quantify the ocean's economic value. They did so by adding up the dollar value of the various benefits gained through fisheries, tourism, shipping, and coastal protection from coral reefs and mangroves, to name a few.

This massive analysis concluded that the sea is worth . . . wait for it . . . US$24 trillion. That's trillion, with a "T." Its annual gross marine product (the equivalent of GDP) was estimated to be US$2.5 trillion a year. That would rank the ocean as the world's seventh largest economy, sitting between the United Kingdom and Brazil. Maybe, just maybe the ocean should have a seat at the G7 Summit . . . we're just sayin'.

Here's the breakdown from the study:

Ocean Asset	US$ Trillion
Marine fisheries	2.9
Mangroves	1.0
Coral reefs	0.9
Seagrass	2.1
Shipping lanes	5.2
Productive coastline	7.8
Carbon absorption	4.3
TOTAL	24

Back to the study. The people crunching the numbers admitted to being unable to factor in less tangible benefits of the ocean — you know, little things like producing over half of Earth's oxygen, anchoring the water cycle, and regulating weather and climate. They didn't even consider oil and minerals (from offshore drilling and mining), wind power, or the ocean as a source of new medicines. As a result, their estimates represent a vast undervaluation of the ocean's worth to humanity, let alone all the other living creatures on the planet, but it is a good place to start and underscores just how important the ocean is to our global economy.

Visit `www.worldwildlife.org/publications/reviving-the-oceans-economy-the-case-for-action-2015` for additional details from the study.

Serving as a Source of Mystery and Wonder

Since the beginning of time, people have been seduced by the sea, its sound, its power, its vastness, and its creatures, all of which have captivated, enchanted, and inspired humankind. The ocean has served for many generations as muse to some of the world's greatest authors, inspiring classics such as *The Old Man & The Sea, Moby Dick, Twenty Thousand Leagues Under the Sea, Treasure Island, Homer's Odyssey, The Tempest,* and more. Artists have drawn, painted, and sculpted its seascape and creatures, both real and mythical. Musicians have sung of the sea's soul for millennia. And the ocean plays a prominent role in many religions across the globe.

While it continues to inspire authors and artists, more recently in human history, the ocean has also inspired scientists, and it continues to do so. Exploring the ocean from surface to seafloor and from coastlines to the middle of the deep blue sea not only goes a long way toward satisfying human curiosity, but also uncovers new sources of food, medicines, and energy; provides insight on how to conserve ocean resources; and enhances safety through a better understanding of earthquakes, tsunamis, hurricanes, and other natural threats to human life and property. In addition, the physical challenges of deep-sea exploration often lead to valuable innovations.

In this section, we present some of the ways the ocean feeds the imagination and a few of the more practical benefits of ocean exploration.

Stimulating our imaginations

People always have been and continue to be fascinated with the bizarre, massive, and powerful creatures lurking beneath the ocean's surface. Before people could explore the underwater world for themselves, they imagined great beasts roaming the ocean, such as the kraken — an enormous octopus-like creature that sailors thought could crush wooden ships whole. Don't worry, nobody's going to "release the kraken"; it's a mythological creature. However, the kraken was probably inspired by a real sea monster — the giant squid, which prowls the deep sea. Females can grow up to 13 meters (43 feet) long (about the length of a school bus) and we think it's fair to say that a squid of that size would scare the sandals off anyone aboard a boat, wooden or otherwise.

Another favorite marine creature of lore is the mermaid. Half human, half fish, and totally voluptuous, these beautiful sirens would sing out to ships and lure men to love or certain death. Mermaids aren't real (though I, Ashlan, still love to pretend I am one while scuba diving). They're believed to have been dreamt up by a bunch of men who were stuck at sea for months and months on end. On the high seas back in the day, women were scarce, and ships were full of dudes with nothing to occupy their free time but rum, stale bread, and tall tales. Historians agree that sailors very likely did see something, maybe a manatee, a dugong, or a sexy seal, but certainly not a mermaid or a lady in a wetsuit. However, after a few glasses of said rum, to a lonely desperate sailor peering through half sober eyes, a manatee might look like a lovely plump lady-fish.

Of course, a few mind-blowing creatures really have existed, some of which were around long before sailors and pirates were day dreaming of love affairs with mermaids. The megalodon, for example, was an enormous ancient shark whose name literally means "giant tooth." Alive during the Miocene and Pliocene epochs (23 million to 2.6 million years ago), megalodons were the largest fish to swim the seas, at least the largest we know of to date. A single meg tooth is the size of an average human hand! Based on scientists' belief that megalodons were almost the same shape as our modern great white shark and based on the size of the teeth and vertebrae fossils found, researchers estimate that this massive shark could grow to be 82 feet long, with a jaw large enough for an adult-sized human to walk through. In comparison, the fictional great white shark in *Jaws* was a mere 25 feet long. But notice that we've been using past tense for megalodons. Yes, they're extinct. We promise.

Increasing our knowledge of the world around us

Unfortunately, in the world of science, ocean explorers often take a back seat to astronauts. Popular movies and the media often depict marine biologists as nerdy bookworms or beach bums spending as much time working on their tans as they do researching the ocean. (Admittedly a tan can be a perk of the job) Astronauts, on the other hand, are always portrayed as brave pilots who train vigorously and risk their lives to explore "the final frontier."

Admittedly, we'd rather be on a beach or a boat than packed like sardines into a space capsule or walking around on a dusty gray moon, but we assure you that ocean explorers are just as serious and committed to their work, and that it can be more dangerous. We can also assure you that the work of ocean explorers is just as valuable as that of space explorers, if not more so. Ask yourself if knowing whether or not water on Mars matters at all to your existence (hint: it doesn't). Okay, okay, researching the history of water on Mars is valuable to understanding other planets and even our own to a certain extent. But knowing and protecting our ocean is critical to our very SURVIVAL on this planet.

Don't get us wrong, we think space exploration is cool. The problem is that the U.S. government invests at least 150 times more money in exploring space than it does exploring the ocean. That would be fine if we had endless money, but we don't. In a world with dwindling resources and a growing population struggling with climate change, maybe sending more money into space isn't the best allo-cation of resources. "But wait — can't we have both you might ask?" Sure, that would be great! But, to have both, equally, would mean that the oceans should get at least the same budget as space, and that is far from the case.

Here are just a few of the many practical benefits of ocean exploration:

>> As resources become scarcer, people are having to go deeper to get them. Ocean exploration leads to the discovery of new sources of food, energy, and other resources and provides a better understanding of how to safely and sustainably tap the ocean for these resources.

>> Increased knowledge and understanding of conditions in the deep sea better prepare humans to respond in the event of deep-sea disasters, such as sunken ships, lost submarines, and offshore oil drilling accidents.

>> Deep-sea exploration led to the 1977 discovery of rich ecosystems that devel-oped around deep-sea *hydrothermal vents* — cracks in the ocean floor out of which percolates magma-heated water chock full of chemicals. Prior to this discovery, scientists believed only sunlight could provide the energy required to grow enough food to support large, diverse ecosystems. But here, at the bottom

of the sea, where the sun doesn't shine, microbes evolved the capacity to extract energy from chemicals through a process called *chemosynthesis*. This discovery completely rewrote the definition of "life" and the scientific understanding of what was required to support it. (See Chapter 5 for more about ecosystems that develop around hydrothermal vents and how it's possible.)

>> Various technologies used for exploring and monitoring the ocean continuously gather data and make it available for analysis to help predict earthquakes, tsunamis, and hurricanes.

>> Ocean exploration collects and analyzes valuable data related to the health and abundance of different marine species, which is crucial for detecting early warning signs of declining populations and revealing possible solutions.

These are just a few of the practical benefits of ocean exploration. Many more will certainly become apparent as ocean exploration evolves. Currently, only about 5 percent of the ocean has been explored, so our understanding of the ocean is still in its infancy. In fact, we know more about the surface of our moon than we know about our ocean, and we have better maps of Mars than we do of the ocean floor. Yet, the future of ocean exploration has some very exciting prospects and much more practical applications. One more thing, for those who think astronauts are cooler than ocean explorers — guess where they do some of their most serious training . . . underwater. That's right, astronauts train underwater to prepare them for space, because surviving in space is easier than surviving underwater. So who's more badass now?

Getting in touch with our emotional connection to the sea

When looking at a map, many see the ocean as a vast expanse between continents, countries, and people — as something that divides us. But the ocean doesn't divide us, it connects us. We are all touched by the sea every minute of every day, no matter where we live. When you look out across the ocean's deep blue hues or envision them through your imagination you're filled with excitement, wonder, curiosity, fear, hope, or a harmony of emotions. As the latest generation of hipsters may say, the ocean fills us with "all the feels."

Have you ever noticed when you're near the ocean, you feel calmer? You feel a little healthier, less stressed, peaceful, and happier? Well, while part of that sense of serenity may be due to the fact that it's the weekend or you're on vacation, most of it is actually due to the fact that the ocean has a soothing and restorative effect on the mind and spirit. This is something that our friend and noted marine biologist Wallace J. Nichols explores in his book *Blue Mind* (Back Bay Books, 2015). In it Nichols explores neuroscience, evolutionary biology, and medical research to illuminate the physiological and brain processes that underlie our connection to water and its effects on the human body, mind, and spirit.

Through his research, Nichols found that being in or close to water, whether ocean, lake, stream, river, or even swimming pool, reduces anxiety, boosts creativity, strengthens human bonds, increases compassion, and even enhances job performance. As Nichols summarizes (www.wallacejnichols.org/122/bluemind.html), "Water is medicine for those who need it most . . . and everyone else," a statement that resonates with Danish writer Isak Dinesen, who wrote, "The cure for anything is salt water: sweat, tears or the sea."

Any way you slice it, we are all connected to the ocean and through the ocean to one another — physically through air and water, emotionally through stories, art and music, and mentally through our daydreams, whether they're dreams of sitting near the beach with a cold drink as the sea blows gently through our hair, of standing tall on the deck of a wooden ship battling the kraken, or of being seduced by the siren songs of distant mermaids.

See you there.

Chapter **3**

Looking Back at the Ocean's History (and Prehistory)

You can understand a great deal about people, places, and things by examining their past. This is even more true of the ocean, which has at least a 3.8 billion-year history, give or take a few hundred million years. Now *that's* a lot of birthdays! Over the course of its existence, it went quickly from mostly fresh water to salt water, has seen entire populations of plants and animals evolve and go extinct, and has been divided into oceans (plural) as huge land masses drifted apart. It is even thought to have frozen over at least twice and possibly as many as four times in its long history. (Don't worry, the last freeze was about 600 million years ago, and the bigger problem now is global warming, *not* cooling.)

In this chapter, we transport you back in time to the birth of the ocean and trace its long history of supporting the evolution of various forms of marine life, a few of which you'll meet up close and personal. We then fast-forward to the present to describe the current condition of the ocean and its inhabitants, along with the impact it has on Earth overall. We wrap things up by taking a peek into the possible future of the ocean to see where it may be heading. (This chapter gets wet, messy, and maybe a little hard to follow at times, so if you want to skip ahead, we would understand. But it's also really cool, so we hope you stick with us.)

Discovering How the Ocean Got Its Start

Our ocean covers 71 percent of our planet and accounts for nearly 97 percent of its water. That may represent a mere drop in a bucket on a cosmic scale, but it's respectable on a planetary scale.

If you had that much water in your basement, you'd want to know where it came from, and scientists have been trying to answer that question since, well, about the time they started asking questions. The two big questions are: *Was that water always here?* and if not, *How the heck did it get here?* These aren't exactly "Which came first, the chicken or the egg?" questions, but they're sort of along the same lines. Either water came from the space debris that formed Earth (so the makings of the ocean were here already), or water arrived via comets or asteroids crashing into Earth after it was already formed. (**Note:** *Comets* are made of dust, rock, and ice, and they tend to fly farther from the sun than asteroids do. *Asteroids* are made mostly of metal and rock and varying amounts of water and tend to hang out closer to the sun. *Meteors* are comets that enter Earth's atmosphere.)

In this section, we present the two leading theories of how all that water got here as we explore the ocean's formation.

REMEMBER

Some scientists prefer one theory over the other, but most believe Earth's water came from multiple sources. Most likely, Earth had some water baked into it during its formation, Earth produced its own water from hydrogen and oxygen, and water was delivered from space via comets and asteroids. That should make everybody happy.

The wet planet theory

The prevailing theory is that the water and/or the chemicals needed to make water were already here when Earth was formed. In other words, Earth formed as a "wet planet." How planets are formed is also a subject of debate, but generally speaking, they form when particles of dust and gas clump together. In its early days, our solar system was a cloud of dust and gas (or clumps of dust and gas). Gravity caused the matter to collapse in on itself as it began to spin, forming the Sun at the center and the planets around it.

According to this theory, the ocean was formed when water (in the form of vapor) slowly escaped from Earth's hot molten interior into the atmosphere surrounding the cooling planet. This *degassing*, as it's called, occurred over millions of years. As the planet cooled to below the boiling point of water, the vapor slowly condensed into clouds and rain began to fall for centuries or even millennia. At some point, estimated at between 4.4 to 3.8 billion years ago, enough water had been wrung from the sky to create the primeval ocean.

The water delivery truck theory

The competing theory is that water came after Earth was fully formed. This theory asserts that thousands of watery/icy comets and asteroids containing hydroxide (a water precursor) delivered water to Earth. Back then, Jupiter was slightly closer to the sun, and its presence and gravity could have shifted the orbit of these comets and asteroids, putting them on a collision course with Earth. Heat from Earth and the sun melted the ice and formed the ocean.

To test whether Earth's water came from comets or asteroids, scientists looked at the composition of the water itself. They compared the ratios of hydrogen isotopes (hydrogen atoms with slightly different nuclear masses) among water samples from Earth, asteroids, and comets. None of them matched exactly, but Earth's water was more similar to that of water contained in asteroids.

Tracing the Evolution of Ocean Life

Although scientists may not be able to pin down where all the ocean water came from, one thing we do know for sure is that life on Earth started in the ocean. In fact, it started not too long after the ocean was formed and has continued evolving ever since. Although the ocean probably existed for at least a few hundred million years before signs of life appeared, that's not very long from a geological perspective. Because life has existed in the ocean for most of the time that the ocean itself has been around, most of the ocean's history is commonly presented as a timeline designating key stages in Earth's geological progression coupled with the corresponding evolution of life. Kind of convenient, isn't it? In this section, we trace the fascinating progression of that evolution, but first, we need to define a few key terms:

>> *Life* is the condition that distinguishes plants and animals (organic) from inorganic matter, such as rocks, minerals, metals, and other substances not derived from living organisms.

>> *Evolution* is the process by which species of organisms arise from earlier life-forms and change over time through natural selection.

>> *Natural selection* is the process by which organisms better adapted to their environment tend to survive and produce more offspring, while those less adapted tend to die out.

Getting the evolutionary ball rolling

So how did life begin? Well, honestly we still don't know for sure. One theory is *abiogenesis* — that life spontaneously arose from non-living material. Another theory, called *panspermia,* suggests that life came from space on a comet or asteroid. But the leading theory is the *RNA World Hypothesis.* RNA is similar to DNA but is structured as a single strand as opposed to a double strand and is made up of different *nucleobases* (the molecular building blocks of RNA and DNA). According to the RNA World Hypothesis, early life on Earth originated with simple RNA molecules that were able to *self-replicate* (create copies of themselves) and create protein molecules — organic compounds that are an essential part of all living organisms.

Another quality of RNA that makes many scientists believe that RNA drove early evolution is that it has the capacity to evolve through interactions with its environment. The RNA World theory has it that diverse RNA molecules formed (how this happened is still not known) and began to evolve and compete for survival. As they evolved, some RNA molecules began to cooperate with one another to develop genetic code, form proteins, and build cells. Eventually, RNA gave rise to DNA, which has the capacity to store more complex blueprints for living things.

Going cellular

However life started, for it to become more complex, cells were required. (A *cell* is a membrane-bound entity containing molecules to sustain life.) The first cell is thought to have consisted of a membrane composed of phospholipids surrounding self-replicating RNA. *Phospholipids* are fatty acids consisting of two hydrophobic (water-fearing) tails and a hydrophilic (water-loving) head that function as the building blocks of all biological membranes. When placed in water, phospholipids naturally aggregate with their heads facing out and tails facing in, forming a two-layer barrier. A cell membrane composed of phospholipids functions like the "skin" of a cell, separating its contents from what's outside it. In early cells, the membrane functioned as an enclosure for the RNA and other molecules, enabling them to operate as a unit with the capacity to reproduce and evolve.

The first true single-cell organisms to enter the picture were the *prokaryotes* — bacteria and archaea. The distinguishing characteristic of prokaryotes is that they lack a membrane-bound *nucleus* (control center), *mitochondria* (power plant), or other membrane-bound *organelles* (organized structures within a cell). Early prokaryotes are thought to have been *chemoautotrophs* — creating their own energy by oxidizing inorganic compounds. Later, approximately 3.5 billion years ago, cyanobacteria evolved, deriving their energy from *photosynthesis* — using sunlight to synthesize foods from carbon dioxide and water.

Even later, about 1.8 billion years ago, more complex single-cell organisms called eukaryotes appeared. A *eukaryote* is any organism consisting of one or more cells containing membrane-bound organelles including a distinct nucleus that contains DNA in the form of chromosomes. (You still with us?) Eukaryotes include all living organisms except prokaryotes. In other words, you're a eukaryote and probably didn't even know it. However, you do have numerous prokaryotes living on and inside you, most of which are beneficial, and some of which perform essential functions (such as aiding in digestion and synthesizing vitamins your body needs but doesn't get in its diet or produce on its own).

And now for a word about metabolism

Before moving on to the evolution of more complex organisms, we'd like to give a shout out to energy — the power that sustains life and drives evolution. So where does all this energy come from? It comes from a set of life-sustaining chemical reactions in organisms collectively referred to as *metabolism*. These chemical reactions can be divided into two types:

>> *Anabolic* processes build molecules. When you're pumping iron at the gym, anabolic processes are at work synthesizing protein molecules to build muscle. Energy is required to fuel anabolic processes.

>> *Catabolic* processes break down molecules into smaller units, often releasing energy; for example, your body can break down sugar or fat molecules to give you the energy to pump that iron.

All living things use the stuff around them to obtain the energy and molecules they need to carry out vital cellular processes, to reproduce, and, in some cases, to move around. However, every known ecosystem on Earth is fueled by organisms that rely on one of the following two metabolic mechanisms:

>> *Photosynthesis is the best known of these processes and* uses energy from the sun to convert carbon dioxide and water into chemical energy and organic molecules needed for growth. Lucky for us, oxygen is released as a waste product

>> *Chemosynthesis is less well known and* uses energy stored in the chemical bonds of inorganic chemicals such as hydrogen sulfide and methane to make glucose from carbon dioxide and water. Chemosynthesis is what enables bacteria to live near hydrothermal vents at the bottom of the deep blue sea. (See Chapter 5 for more about life that exists around hydrothermal vents.)

REMEMBER

Organisms that rely on chemosynthesis and photosynthesis anchor the food webs that enable ecosystems to develop. They're sort of like the farmers who grow all the food, except they are the food!

All together now: Multicellular organisms

Over time, cells began to gather and hang out together, probably not out of loneliness but because sticking together was advantageous to each cell in the group. It's sort of like schools of fish forming to ward off predators or the way some plants and animals form symbiotic relationships; for example, a sea anemone's tentacles protect a clownfish from predators while the clownfish chases away butterfly fish that would eat the anemone. Of course, symbiosis is different at the cellular level, but the concept is the same.

For whatever reason, cells began to aggregate forming filaments or mats consisting of the *same* cell types (colonies) or *different* cell types (symbiosis). Over time, cells formed clumps and then the clumps formed more and more intricate structures with different parts of each structure performing a distinct function; for example, cells at one end of the structure could be in charge of consuming nutrients, while cells at the other end could be in charge of eliminating waste products.

How the first multicellular organisms developed and then how more complex organisms developed with distinct organs and limbs are topics of speculation. What we do know is that the first multicellular organisms arrived on stage — about 600 million years ago, which is relatively recent in the ocean's 3.8 billion-year history.

Multicellular life really took off in the Ediacaran period (from 635 to 541 million years ago) with simple organisms such as branching rangeomorphs (animals shaped like leaves), the kimberella (sort of like a slug), and the spriggina (similar to a trilobite; see Figure 3-1), as well as early sponges and cnidarians (jellyfish, anemones, and so on), and soft-bodied organisms that looked like worms, corals, sea-pens, seaweed, and lichen. Dickinsonia is another famous organism from this period, and may be the first animal to move on its own, scampering across the seafloor some 567 to 550 million years ago. Some of these may be the first *metazoans* — animals with specialized cells and different body sections for different roles.

FIGURE 3-1:
Trilobite fossil.

Taking evolution to the next level in the Paleozoic era

Buckle up, folks . . . this is where things start to get really interesting. Don't be turned off by the hard-to-pronounce words or events that can be a little confusing (we had to study this many times before it sunk in). This section begins the story of the evolution of complex life-forms on Earth, and it is a fascinating one. If anything, it reminds us of the incredible dynamism and complex wonder that resulted in . . . well . . . us. So read on and marvel at just how amazing this world really is.

The Paleozoic era spans from about 541 to 251 million years ago, when life underwent enormous diversification. (*Paleozoic* roughly translates to "ancient life.") It began with the *Cambrian explosion* (when nearly all major animal phyla appeared) and ended with *The Great Dying* (a mass extinction) and can be subdivided into the Cambrian, Ordovician, Silurian, Devonian, Carboniferous, and Permian periods.

The Cambrian period

The Cambrian period (from 541 to 485.4 million years ago) was a time of massive diversification of life in the ocean. It was thought to have begun as a result of changing ocean chemistry (due to erosion and minerals washing into the ocean) and a boom in oxygen levels due to growing populations of phytoplankton (see Chapters 7 and 8 for more about phytoplankton). During this period:

» Shells and exoskeletons with new body plans evolved.

» The first complex animals with mineralized remains formed.

» The evolution of flexible limbs became the first "arms."

» The first real predators evolved, such as anomalocaris, a shrimplike creature that hunted worms and other soft bodied animals.

» Pikaia (see Figure 3-2) and haikouella grew a flexible rod of cartilage to swim, becoming the ancestors of the *vertebrates* (animals with backbones).

» Trilobites, one of the first *arthropods* (think spider or crab), appeared.

» More complex food webs began to develop.

This period ended with the Cambrian–Ordovician Extinction event, due to a possible crash in oxygen levels in the ocean.

FIGURE 3-2: Artist rendering of a Pikaia with a jellyfish.

Photo by Nobu Tamura. Licensed under CC BY-SA 2.0

The Ordovician period

The Ordovician period (from 485.4 to 443.8 million years ago) changed the face of the world. Rapid seafloor spreading resulted in high sea levels, creating new environments and habitats as well as a rapid diversification of life, including the Great Ordovician Biodiversification event. During this period:

>> True fish appeared (see Chapter 12), the *ostracoderms* — jawless fish with bony plates.

>> Ocean life also consisted of *graptolites* (which lived in colonies sharing the same skeleton), corals, *crinoids* (think starfish, sea urchins), *brachiopods* (clams, oysters), the surviving trilobites, *gastropods* (snails, slugs), *cephalopods* (squid, octopus), and red and green algae.

>> Also swimming around during this period are six-meter-long shelled cephalo-pods called *cameroceras* (imagine a squid in a long pointy shell, three times longer than you are tall).

>> The first land plants started to grow, resembling moss. These plants sucked up a lot of carbon dioxide helping to create an ice age, which is basically the opposite of what is happening now — global warming as a result of a massive release of greenhouse gasses, including carbon dioxide (thanks to human activities).

>> The Ordovician-Silurian Extinction event occurred, wiping out 86 percent of marine species, including some trilobites and cephalopods. (These periods never seem to end well.)

The Silurian period

The Silurian period (from 443.8 to 419.4 million years ago) was a time of stable climate (well, relative to earlier periods), but warmer, causing sea levels to rise again. So fun. During this period:

>> Early vascular plants appear, and fungi move onto land. This is also possibly when spiders and centipedes show up.

>> Fish split into bony fish and cartilaginous fish. The cartilaginous fish eventually become sharks and rays.

>> The bony fish split into two groups — lobe-finned and ray-finned. The ray-finned fish give rise to modern fish, while the lobe-finned fish evolve into *tetrapods* (generally four-limbed creatures), which later evolve into amphib-ians, reptiles, mammals, and birds.

>> The first evidence appears showing coral reefs expanding and developing.

- **Brachiopods** (like oysters) are very common, but other organisms continue to thrive, including trilobites, echinoderms, cephalopods, and gastropods.

- *Ostracoderms* (jawless fish) diversify, and the first freshwater species evolves.

- Sea scorpions called eurypterids (sort of like horseshoe crabs) evolve, and holy moly are they ever terrifying! (See Figure 3-3.)

- Fish get jaws, although it will be about 430 million more years before *Jaws*, the book and movie, become popular entertainment.

FIGURE 3-3: Artist rendition of a sea scorpion (eurypterid).

Photo by Obsidian Soul with background by Dimitris Siskopoulos. Licensed under the CC BY 4.0.

The Silurian period ended in a series of smaller extinction events, likely caused by a drop in sea-level, which the bottom-dwellers couldn't handle; mostly cephalo-pods went extinct.

The Devonian period

The Devonian period (from 419.2 to 358.9 million years ago) marks the era when fish took over the seas. Sea levels were higher, covering much of the land, creating vast shallow coastal areas. During this period:

» Common animals are rugose corals, crinoids, and jawless fish.

» Early sharks emerge (hey girl, hey!)

» *Placoderms* (jawed, armored fish) dominate and occupy both sea and freshwater environments. Some are predators, others are filter feeders.

» Lobe-finned fish relatives develop the ability to move along shorelines (on land), like mudskippers today. 379 million years ago, their descendants finally became land dwellers — *tetrapods* (meaning four legs). The proof? They left footprints.

» Amphibians evolve.

» *Coelacanths* (large, plump, lobe-finned fish) evolve and are still around to this day. See Chapter 12 for a cool photo.

» The first spiders and other insects appear, this time for sure.

» Crinoids, corals, brachiopods, ammonite relatives, and ostracoderms are present and accounted for. *Ammonites* have a shell like a snail and a body like a squid.

» The first complex land ecosystems begin to develop.

The Late Devonian Extinction marked the end of the party, occurring in at least two phases due to drops in oxygen levels in the ocean. Even the trilobites and placoderms had a tough time surviving this mass extinction.

The Carboniferous period

The Carboniferous period (from 358.9 to 298.9 million years ago) was marked by a warm, humid climate with loads of oxygen and lots of forests, making this period famous for its swamps. Some animals called amniotes, began laying eggs with shells on land; these are the early ancestors of reptiles, birds, and mammals. But it wasn't all about them; arthropods (insects) also began to take over the land — think giant dragonflies the size of seagulls.

Meanwhile, back in the ocean:

» Sharks begin their massive diversification with about 45 families emerging.

» Coral reefs flourish again, and invertebrate marine animals are common. *Foraminifera* (single-cell organisms with shells) become common.

» Nautiloids emerge, and while they represented a very diverse group of predators at the time, today only their ancestor the Nautilus remains (see Figure 3-4).

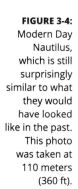

FIGURE 3-4: Modern Day Nautilus, which is still surprisingly similar to what they would have looked like in the past. This photo was taken at 110 meters (360 ft).

Source: Laurent Ballesta – laurentballesta.com

This period ended as the continents started to merge to form the most well-known supercontinent, Pangea. Because inland areas of this massive continent were too far from the ocean to get any moisture, large deserts emerged in Pangea, and forests died out (the *carboniferous rainforest collapse*). The amniotes we mention above (early reptiles) were better able to adapt to this drier climate and continued to diversify, while creatures that had dominated in earlier, wetter periods, such as amphibians, struggled.

The Permian period

Earth's climate continued to get hotter and drier during the Permian period (298.9 to 251.9 million years ago). The seas were mostly still warm and shallow, and ocean life was similar to that of the Carboniferous period. While ammonoids (early marine mollusks) became more complex, not much else really changed. Sponges and corals made reefs, Bryozoa (moss animals) emerged, fish, ammonoids, gastropods, brachiopods, and echinoderms were still common.

All was hunky-dory till about 252 million years ago, when the Permian period ended with "the Great Dying" — Earth's most extreme extinction event *ever*. Ninety-six percent of all marine species and 70 percent of all terrestrial species were wiped off the planet. Talk about a sad ending! This mass extinction event was most likely caused by one or more of the following:

» Climate change due to a mass release of methane into the atmosphere from the oceans

» Volcanic activity/eruptions in Siberia

>> A really big rock slamming into the Earth (a large crater matching the age of this event has been found near the Falkland Islands off the coast of Argentina)

And on that happy note, we come to the end of the Paleozoic era . . . but think about all the good times we had: the diversity of life-forms exploded, plants and animals spread from sea to land, and everyone's favorite super-continent, Pangea, was formed. It also set the stage for . . . wait for it . . .

Gaining momentum in the Mesozoic era

Sorry, are you still waiting? The Mesozoic era (from 252 to 66 million years ago) is known as the Age of Reptiles or, for you terrestrial loving tree-huggers, the Age of the Conifers, not to mention the freaking DINOSAURS!!!!! This era is divided into only three periods (phew).

Spoiler Alert: If you want to skip the next few sections, here's the plot summary: Early mammals, birds, amphibians, bees, and flowering plants evolved. Dinosaurs took over the planet (woohoo!), then all non-avian (not birdlike) dinosaurs went extinct (booo!). The Mesozoic marine revolution occurred, during which shell-crushing marine reptiles evolved, causing organisms with shells to grow stronger, spinier shells in self-defense, which sparked an evolutionary arms race that continues to this day.

The Triassic period

The Triassic period (from 251.9 to 201.3 million years ago) was a time of rebuilding. After the Great Dying, few creatures remained. (Talk about a shrinking dating pool!) So, who was left? *Temnospondyls* (early amphibians) and *therapsids* (early relatives of mammals) made the cut, as did some fish, including some small sharks, and some mainly amphibious marine reptiles. Most animals living in shells fared poorly due to ocean acidification, which hinders the ability to form shells (more on that later). Hinged brachiopods, crinoids (sea lilies), and some ammonites (early mollusks) did okay.

During this period of rebuilding:

>> The first *ichthyosaurs* (fishlike reptiles) evolve, probably from terrestrial ancestors that returned to the ocean; for example, the cartorhynchus (say that three times fast) has flippers like a seal and a tail like a lizard. Ichthyosaurs were predators, feeding on fish, shellfish, reptiles, and other ichthyosaurs. Some were ram feeders, like today's whale shark, swimming toward their prey with mouth open to engulf the prey and the water around it. Others could crush shells or bore into them.

>> *Archosaurs* (which means *ruling reptiles*) appeared on the scene. Broadly classified as reptiles, this group includes all extinct dinosaurs along with birds and crocodiles.

>> The archosaurs split into two groups: Saurischia (lizard-hipped) and Ornithischia (bird-hipped).

>> The *plesiosaurs* arrive — large marine reptiles with four flippers, a long neck, and a long tail (see Figure 3-5).

FIGURE 3-5:
An average, everyday plesiosaur.

Source: National Park Service, Public Domain

This period abruptly ceased with the End–Triassic Mass Extinction, also known as the Triassic–Jurassic (T-J) Extinction event. Around the same time, Pangea started to break apart, with North America drifting off first. These were triggered by massive shifts in the tectonic plates which caused an enormous spike in volcanic activity and a resulting increase in greenhouse gasses, which really cranked up the heat. This particular mass extinction caused the loss of the temnospondyls (sort of like salamanders), therapsids (mammal–like reptiles), and most ichthyosaurs (though some larger species survived).

The Jurassic period

The Jurassic period (from 201.3 to 145 million years ago) is everyone's favorite (thank you Michael Crichton and Jeff Goldblum!). Because you've probably seen the movie, we'll keep it simple . . . ish. During this period:

- *Ornithischians* (vegetarian dinosaurs with bird hips) begin to spread and diversify.

- Adaptations to leg and pelvic bones allow for a larger gut and, hence, larger teeth to help fill that gut, ultimately resulting in larger dinosaurs.

- *Megazostrodon*, the first true mammal appears, but unfortunately it's as ugly as today's possums . . . even though they are kinda cute.

- Some mammals become aquatic (such as the castorocauda, a beaverlike critter) and some take to the air (such as the volaticotherium, similar to today's flying squirrels, definitely cute).

- Paravian dinosaurs appear, sporting feathers in place of scales (très chic), becoming, unbeknownst to them, ancestors of modern birds.

- Large marine reptiles rule the seas as main *apex predators* (top of the food web with nothing big enough or mean enough to eat them). These marine reptiles include plesiosaurs and ichthyosaurs (which look like plesiosaurs without necks) and, later, pliosaurs (close cousins of ichthyosaurs) and marine crocodiles.

- Diverse ecosystems develop complete with ammonites, gastropods, and fish.

- Freshwater turtles put in an appearance.

The Jurassic period ended with a relatively minor extinction event (what a disappointment). Pangea continued to go to pieces, and sea levels began to rise, creating shallow seas in North America and Europe. These rising seas coupled with a change in climate caused by an increase in volcanic activity in the Pacific may have been the cause.

The Cretaceous period

Life literally started to bloom during the Cretaceous period (145 to 66 million years ago), during which time flowers first appeared in the form of a genus of plants known as *Archaefructus*. Of course, plenty of other activity was going on as well:

- *Tyrannosaurids* (early tyrannosaurus relatives) evolved.

- *Ceratopsians* (those dinosaurs that look like rhinos) grew their head gear.

- Several awesome dinosaurs stepped on stage, including: Ankylosaurs (sort of a cross between a horned toad and an armadillo), titanosaurs (the reptile version of a giraffe), hadrosaurs (hard to describe), and azhdarchids (giant flying pterosaurs).

- *Mosasaurs* appeared, looking like a cross between a whale and an alligator.

>> Rays and modern sharks became common.

>> *Hesperornithes* (flightless birds that look like penguins but with longer necks) evolve.

Near the end of the Cretaceous period, species diversity was low and things were going to get worse with the Cretaceous-Paleogene (K-Pg) Extinction event, also known as the Cretaceous-Tertiary (K-T) Extinction, probably caused by the famous asteroid that killed all the dinosaurs. As the story goes, a mountain-sized asteroid hit the gulf of Mexico, spewing ash and blocking sunlight, causing a serious cold spell. Gases also affected the land and ocean resulting in the death of plants and collapse of ecosystems. This period is also responsible for some of the largest volcanic eruptions in history, lasting for tens of thousands of years (yikes!). All ichthyosaurs, plesiosaurs, and mosasaurs went extinct. Plankton populations collapsed, causing food webs to disintegrate. All in all, about 75 percent of the world's species were wiped out.

Increasing sophistication in the Cenozoic era

And here we are, the era you've all been waiting for, the Cenozoic era (from 66 million years ago right up to today), the era that marks the rise of the mammals, including modern humans. In this section, we lead you through the Paleogene, Neogene, and the Quaternary periods.

The Paleogene period

So much happened during the Paleogene period that we're going to break it down into its three epochs: the Paleocene, Eocene, and Oligocene.

THE PALEOCENE EPOCH

The Paleocene epoch (from 60 to 55 million years ago) was pretty much devoid of large animals, as small creatures were rapidly evolving to fill empty niches, and the planet warmed so much that rainforests sprouted even at the poles. Here are some of the highlights:

>> *Waimanu* (a flightless water bird) claimed fame as the earliest penguin, while birds experienced a high degree of *speciation* (the evolutionary process by which populations evolve to become distinct species).

>> Mammals also diverged into the modern groups of *monotremes* (mammals that lay eggs), *marsupials* (mammals that carry their babies in pouches), and *eutherians* (mammals with placentas; for example: humans).

- » Early ungulate-like mammals appear (*ungulate* means hooved).

- » *Creodonts* (carnivorous mammals) appear, generally belonging to one of the following two groups: *galecyon* (similar to dogs) and *oxyaena* (more like cats).

- » *Plesiadapiforms* (primate-like mammals that took advantage of a heavily forested Earth) make their appearance.

- » Most amphibians, freshwater crocodiles, and turtles survived the K-Pg Extinction event, but marine life took a long time to recover. The exception were small pelagic fish, which recovered quickly.

- » Ancestors of the megalodons (the huge sharks we introduced you to in Chapter 2) appeared, and ray-finned fish took over the oceans.

The Paleocene epoch ended with the Paleocene–Eocene thermal maximum about 55 million years ago, marked by a massive release of carbon into the atmosphere over the course of 20,000 to 50,000 years, which raised the average temperature on land 5 to 8 degrees Celsius (9 to 14.4 degrees Fahrenheit). Any way you measure it, that's huge in terms of global warming, and the warming lasted about 200,000 years. The huge carbon surplus also caused ocean acidification, which killed a lot of marine species. (Acidification currently threatens ocean life, as explained in Chapter 21).

THE EOCENE EPOCH

During the Eocene epoch (from 55 to 33.9 million years ago), Earth suddenly went from greenhouse to icehouse. One of the proposed causes of this dramatic change in climate is *the Azolla event*, during which blooms of azolla freshwater ferns in the Arctic Ocean sucked massive amounts of carbon dioxide out of the atmosphere. During this epoch:

- » The ice sheets expanded.

- » Deciduous trees, which drop their leaves in the fall, became favored over evergreens.

- » Early *perissodactyls* (odd-toed hoofed animals), *artiodactyls* (even-toed hoofed animals), and primates appear on land.

- » In the oceans, *basilosaurids* and *ambulocetus* (both ancestors of modern whales) appear, as well as the ancestor of *sirenians* (the politically correct name for sea cows such as manatees).

- » Most modern mammal orders appear, including bats, rodents, and *proboscideans* (mammals with long snouts such as wooly mammoths and elephants).

- » Giant snakes, such as the titanoboa, and giant flightless birds also make their entrance.

The Eocene ended with the Eocene-Oligocene Extinction event. This was a period of cooling, which may have been caused by another meteorite impact, causing a large turnover of mammalian and aquatic species in particular. During this time the *archaeoceti* (primitive whales) went extinct.

THE OLIGOCENE EPOCH

During the Oligocene epoch (from 33.9 to 23 million years ago), ice sheets began to form again at the poles, which lowered sea levels, and more recognizable life-forms emerged:

>> Grasslands flourished on land, making hoofed herbivores (such as rhinoceros and horses) more common, and *ruminants* (like cows) evolved to gobble up the grass.

>> Big cats, dogs, horses, camels, eagles, raptors, elephants, and deer appear.

>> *Mastodons* (the elephant's nearly identical twin) and *Paraceratherium,* the largest land mammal of all time, wander the earth.

>> Old world monkeys split from new world monkeys, and our ape ancestors enter the picture.

>> Marine life mainly resembles what it is today, but as a whole, marine life diversity declines. Even so, during the Oligocene epoch, the ocean is home to baleen and toothed whales, *desmostylians* (vegetarian sea rhinos), and *pinnipeds* (seals, sea lions, and walruses).

No major event signified the end of the Oligocene epoch; it's marked by small changes in fossils of algae and *foraminifera* (microscopic marine organisms). Nice to not have a mass extinction for a change, right?

The Neogene period

The Neogene period (from 23 to 2.6 million years ago) is divided into the Miocene and Pliocene epochs, but don't worry we're not going to slice and dice this period. Here are some of the highlights:

>> Desmostylia (that crazy sea rhino thing) goes extinct.

>> Kelp makes its debut by forming its own kelp forests.

>> Aquatic sloths and sea otters appear.

>> *Cetaceans* (whales and dolphins) become more diverse in correlation with the increase of giant marine predators such as megatooth sharks and predatory sperm whales.

>> The continents are pretty much where they are today. North and South America are connected with a land bridge, enabling the *Great American Biotic Interchange* — when species moved freely between the two continents (open borders anyone?).

>> Plant and animal species begin to take on a more modern appearance, and most modern bird groups are established.

>> The first apes evolve from old world monkeys.

>> Early human ancestors, such as *Australopithecus,* introduce themselves and demonstrate their ability to create and use stone tools (Stone axes? Awesome).

>> The megalodon arrives. (We're not afraid of sharks, but we wouldn't want to swim with a megalodon even if we were holding a stone axe.)

>> Our human ancestors from the genus *Homo* arrive.

This period ends with the beginning of an ice age — a very long cold spell characterized by the expansion of continental and polar ice sheets and alpine glaciers.

The Quaternary period

The Quaternary period (from 2.6 million years ago to now) is divided into the Pleistocene and the Holocene epochs, and this time, we're going to honor that division. (Note that some people believe the Quaternary epoch should be part of the Neogene period, but we're going to keep it separate.)

The Pleistocene epoch ushered in some of the most awesome *megafauna* (large animals), including mammoths, mastodons, saber-toothed cats, giant sloths (they would have been so cute and s-o-o-o s-l-o-w), *Megalania* (a lizard you wouldn't want to cross paths with), elephant birds, dire wolves (yes, *Game of Thrones!*), cave bears, wooly rhinoceros, and *aurochs* (wild ox), but most went extinct during the transition to the Holocene epoch. And we also lost megalodons (so, no worries about bumping into an 82-foot-long shark).

Neanderthals evolved during the Pleistocene epoch, alongside more anatomically-modern human ancestors. The oldest living things today are two *nematodes* (wormy looking creatures) from the Pleistocene, who were frozen in ice 42,000 years ago, and were brought back to life, which is creepy and super cool at the same time. And human ancestors became *Homo sapiens* (let's hear it for the humans!). During the Holocene, favorable environmental conditions allowed loads of species to grow and flourish around the globe. The Holocene also marks the beginning of agriculture.

And that pretty much wraps up 3.8 billion years of evolution, not that it has stopped or anything like that.

Taking the Earth's Present Evolutionary Pulse

The evolution of life on Earth reveals three key takeaways: First, life as we know it is doomed. One of these days, BOOM!, massive extinction. Guaranteed. We don't know how or when, but it's gonna happen. Second, the living Earth is very resilient. Barring some cataclysmic cosmic event, such as the sun expanding and gobbling up the planet, life will rebound after the next mass extinction. It may not look anything like life on the planet now, and humans may not make the cut, but that's life! Third, we're living in the midst of a wonderful epoch with a very impressive diversification of species — from microscopic bacteria to coral reefs to the blue whale, from algae to redwoods, from mice to elephants. The air, land, and sea are teeming with life.

Unfortunately, we (the people) are ruining it for ourselves, and we (the authors) aren't going to sugarcoat it. Many scientists suspect we're already entering the sixth mass extinction event on our planet (the five previous ones were mentioned earlier in the section "Tracing the Evolution of Ocean Life"). But this mass extinction, called the Holocene or Anthropocene Extinction, is different from the others, in that no sudden and dramatic natural disaster, like a meteor the size of Manhattan slamming into Earth or a massive volcanic eruption, is responsible. Neither is it being caused by some millennia-long change in the atmosphere. No, this mass extinction is being brought on at a rate unseen in Earth's history, and it is entirely our fault.

According to "The Living Planet Report" published by the World Wildlife Fund (WWF) in 2018, humanity is responsible for at least a 60 percent decline in the population size of all mammals, birds, fish, amphibians, and reptiles since 1970. Wildlife populations in rivers and lakes have declined by 83 percent due to destructive agriculture and dams. And it's estimated that we have lost 50 percent of the total biodiversity on Earth, all in the last 40 years.

As if that's not bad enough, scientists estimate that the average baseline (historical) rate of extinction is one species per one million per year, not counting species lost due to human activities. When you factor in extinctions that can be blamed on humans, that number is hundreds to thousands of times higher. In other words, the world loses thousands of species *every single year* thanks to us!

Stats and reports are easy to gloss over. What drives the point home is personal observation. For example, there was a time when the fishing in the Florida Keys was legendary, with people flocking to the water to haul in a big fish and pose for a photo. While they still do it today, it was long known that the size and diversity of fish was declining dramatically, but it wasn't until 2007 when an intrepid graduate student named Loren McClenachan found an ingenious way to calculate how precipitous that decline actually was. While visiting a local library, Loren discovered a collection of historical photos documenting sportfishing catches over 60 years from the same area. She realized rather quickly that she had discovered something significant. Using a time sequence of dozens and dozens of photos, three of which are here (see Figures 3-6, 3-7, and 3-8), Loren was able to calculate the change in size by measuring over 1,200 fish in the historical pictures. McClenachan concluded that the average fish caught in the 1950s were over 6 feet long, while in 2007, the largest fish were about 1 foot long. That is an 88 percent drop in catch size! Her research was published in the scientific paper: McClenachan, L. 2009, "Documenting loss of large trophy fish from the Florida Keys with historical photographs," *Conservation Biology* 23:636-643.

What recreational fisherman were happy to catch in 2007 would have been laughed at by the fisherman in the 1950s as bait. This is known as *shifting baselines* (see the nearby sidebar for more). People don't know what they're missing out on because their perspective doesn't benefit from historical knowledge. The good news is that aggressive conservation efforts have seen some fish stocks begin to recover over the last few years, but a tremendous amount of work is still needed.

FIGURE 3-6: Fish catch in the Florida Keys 1957.

Source: Monroe County Public Library. Photo by Wil-Art Studio. Licensed under CC BY 2.0

FIGURE 3-7:
Fish catch in
the Florida
Keys 1980s.

FIGURE 3-8:
Fish catch in
the Florida
Keys 2007.

SHIFTING BASELINES

We humans tend to lose perspective over time as we gradually become accustomed to change. For example, as you age, you probably don't notice the little differences from year to year. Then, for some strange reason, you look at a photo of yourself 20 years ago and, crikey, what happened?!

In statistics, this phenomenon is referred to as a *shifting baseline*, which involves comparing something to a reference point (baseline) which may already represent a significant change from a previous reference point. With a shifting baseline, the condition of the ocean doesn't look so bad when you return to the same beach year after year.

Instead of comparing ocean life to that of 10 or 20 years ago, compare it to a truly untouched and healthy ecosystem. Building on the beach example, instead of comparing it to the year before, compare it to what it was a few decades ago, a hundred years ago, what it looked like before the Industrial Revolution, and what it looked like before us.

This lack of historical perspective presents a problem for conservation because it causes people to overlook the massive devastation to fragile ecosystems caused by gradual changes over decades. That's why research and perspective are so important and why books like this one exist — to help provide perspective so we can make well-informed decisions about what we do that impacts the wonderful world around us.

Looking Ahead: What's in Store for the Ocean's Future?

According to the ancient Greek philosopher Heraclitus, "The only constant in life is change," and this holds true for the ocean.

First, let's look at the physical shape of the ocean. The ocean is continually changing shape, albeit at a v-e-r-y s-l-o-w pace. For example, thanks to *plate tectonics* — the movement of the seven humongous slabs of rock that form Earth's crust as we know it (see Chapter 6 for details), the Atlantic Ocean is spreading apart, and the Pacific Ocean is shrinking. Or consider that the Baja Peninsula off the coast of Mexico is slowly "unzipping" from the mainland, and the Red Sea is expanding as the African and the Middle Eastern tectonic plates part company. Both of these movements are happening at about the same rate that your fingernails grow.

And at some time in the near future (like 50 to 200 million years from now), a new supercontinent will form again as the continents smoosh (a totally scientific term) back together. It's already been named — Amasia. Scientist are debating where the supercontinent will actually settle, but wherever it lands, it looks as though the continents will be reunited, and it feels so good.

Now on to the depressing stuff.

If humans continue conducting business as usual, and burning fossil fuels at the current rate, Earth will continue to warm, which means the ocean's temperature will steadily increase, not to mention that the excess carbon in the atmosphere will result in a decrease in pH in the surface ocean, resulting in what is known as ocean acidification. This combination of factors is already resulting in rising sea levels, shifting currents, extreme weather, and a collapse of biodiversity. Just as humans can survive a fever a few degrees higher than normal but die at temperatures over about 106.7 degrees Fahrenheit, so too can most ocean life tolerate an ocean that warms by a few degrees but when it becomes just a little warmer than that, ecosystems collapse, species migrate or just die out and the entire balance of the ocean is ruined.

All of this is BAD for us! Many coastal cities around the world are in the direct path of severe flooding as a warming ocean means rising sea levels. In terms of fishing, we're currently on a path to eating our way down the food chain, and the lower you go, the more disruptive that fishing is, because we humans end up outcompeting the fish for the food *they* need to live. It's not a stretch to say that we're at risk of fishing ourselves out of fish.

Ocean acidification is another huge threat. The ocean is facing potential catastrophic damage from a change in pH (the measure of hydrogen ion concentration in seawater). Before the Industrial Revolution (that is, before humans began to burn fossil fuels), the average pH in the surface ocean was 8.2. In today's ocean, the pH of surface water has dropped below 8.1 because of increased carbon dioxide produced by burning coal, oil, and natural gas. If global warming can be held to 2 degrees Celsius (3.6 degrees Fahrenheit) above the pre-industrial temperature, the average pH will drop to 7.9. A drop of 0.3 pH units might not sound very important, but that is not the case. pH is based on a logarithmic scale (like the Richter scale used to measure the magnitude of earthquakes). A decrease of 0.3 pH units is equivalent to a 45 percent increase in hydrogen ion concentration. If global warming continues to 4 degrees Celsius (7.2 degrees Fahrenheit), which some scientists have argued may occur by 2100 if humanity fails to take any action to achieve net zero carbon emissions by 2050 the average pH of surface ocean water could drop to 7.6. Although seawater at a pH value of 7.6 is still alkaline (seawater pH would have to decrease below 7 to be considered acid), the hydrogen ion concentration would be 280 percent higher than it was before the Industrial

Revolution. This would have a devastating effect on many forms of marine life that are adapted to seawater with higher pH (lower hydrogen concentration). This would not be good, like Doomsday not good.

A warming ocean is also disrupting weather because the ocean plays a huge role in controlling the weather. Those disruptions are already beginning to have an impact on agriculture and the availability of fresh water, which is kind of important because you know . . . food and water. Already, people around the world are suffering from major shifts in temperature and rainfall that are wreaking havoc on crops and drinking water supplies.

Crap.

On the flip side, the ocean is incredibly resilient, and humans are smart, not that we always act on what we know, but we are Earth's most innovative species. We have many tools that we know work to help the seas recover and thrive. One of the best tools is the establishment of Marine Protected Areas (MPAs), and there's a major global push to protect 30 percent of our ocean by 2030 called 30x30. Although these MPAs may be impacted by warming water and the lowering of pH, setting aside safe places for nature has been proven to help stem biodiversity loss and build resilience in nature.

In addition, better fisheries management can restore our fisheries to healthy sustainable levels — enough to feed us while also supporting diverse marine populations.

Our greatest ally in the fight to protect the ocean is the ocean itself. In fact, the ocean already has slowed global warming from greenhouse gasses by absorbing much of the heat and pulling in heat-trapping carbon dioxide from the atmosphere. By restoring critical ocean habitats such as mangrove forests, seagrass meadows, and marshes (see Chapter 21), we can create natural systems that will help us reduce excess carbon in the atmosphere.

REMEMBER

Some people used to think that the ocean was too big to fix, but as former NOAA director Jane Lubchenco says, "It's time to stop thinking of the ocean as a victim of climate change and start thinking of it as a powerful part of the solution." Here are some of the ways the ocean can help reduce global warming:

>> The ocean can serve as a power source to reduce our reliance on fossil fuels. People are already harvesting energy from wind, waves, and tides. The ocean is also the largest collector of solar energy, and through ocean thermal energy conversion (OTEC) and other technologies, people are using that heat to produce electricity. (See Chapter 19 for more about tapping the ocean for energy.)

>> The ocean is the biggest heat sink and carbon sink on the planet, meaning the ocean absorbs a huge amount of the heat and carbon before they significantly impact life on Earth. If we maintain and in some cases restore a healthy blue carbon ecosystem, the ocean can continue to sequester and store carbon, doing much of the heavy lifting required to reduce carbon in the atmosphere.

>> *Sustainable aquaculture* (land-based aquatic farming) and *mariculture* (ocean-based aquatic farming) of both marine plants and animals can help grow food, establish new habitats create new jobs, and absorb some extra carbon as well.

The good news is that by working together in partnership with the ocean, there is the very real possibility that we can stop the downward spiral we have set ourselves upon, restore the ocean to abundance, and rebuild a thriving future for all living creatures both above and below the ocean's surface.

2

Finding Your Way Around

Explore different ways to divide the ocean into zones based on depth, light, distance from shore, and other factors . . . and understand why we divide the ocean into zones.

Grasp the concept of marine *ecosystems* — biological "neighborhoods" where organisms have evolved over time in response to their interactions with their environment and with one another. (Well, now you've grasped it.)

Visit the most popular ocean ecosystems from mudflats, estuaries, and mangrove forests along the coasts to offshore kelp beds and coral reefs to ecosystems that form around hydrothermal vents and cold seeps in the deep ocean.

Explore the geophysics of the ocean — the contours of the ocean floor, how it was shaped over time, what's on and below the seafloor . . . those sorts of things.

Chapter **4**

Mapping the Ocean by Zones

People like to divide really big things into smaller parts to better understand, discuss, and manage them. For instance, your city or town probably has residential, commercial, and maybe even industrial and agricultural zones. Meteorologists divide the world into climate zones — tropical, temperate, arid, continental, and polar. To deliver mail more efficiently, the U.S. Postal Service breaks down the entire country into postal zones, each with its own ZIP code.

Likewise, oceanographers divide the ocean into several zones to better understand the conditions required for different organisms and ecosystems to evolve and exist. They use several criteria to designate a zone, including exposure to tides and currents, the area's topography (physical characteristics), depth, amount of light, and more.

In this chapter, we lead you on an exploration of the different ocean zones. Understanding these zones will enhance your understanding of the ocean and its fascinating habitats and inhabitants . . . assuming you don't zone out while reading this chapter (sorry, we couldn't resist).

Dividing the Ocean into Three Horizontal Zones

The ocean's horizontal zones extend from shoreline to the deep blue sea and are generally classified as shallow, deep, and deeper or, more scientifically, as intertidal, neritic, and oceanic (see Figure 4-1). In this section, we describe each zone, the conditions that make each one special, and some of the more interesting inhabitants that call them home.

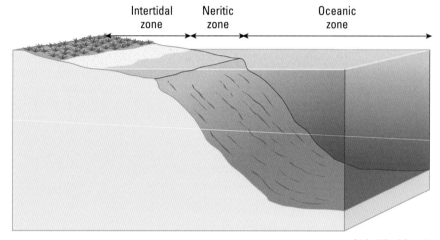

Intertidal zone   Neritic zone   Oceanic zone

©John Wiley & Sons, Inc.

FIGURE 4-1: The ocean's horizontal zones.

FUN FACT

Mobile animals, such as crabs and sea stars, can move between zones to find optimal locations at different times of day, and can even take refuge in wet rock crevices if conditions get too hot and dry.

Where land meets sea: The intertidal zone

Time to dip our toes into the ocean, literally, because if your toes ARE close to shore, you're in the *intertidal zone* (also known as the *littoral zone*) — the area along the coastline that's underwater at high tide and exposed at low tide. Depending on where on the coast you're standing, the intertidal zone can look drastically different. You could be looking at a sandy beach, a mudflat, a rocky shoreline, a marsh, or a mangrove forest, for example.

The intertidal zone can be further divided into low, middle, and high regions, based on how far up from the ocean it is, which generally determines how wet it gets and therefore the nature of the marine life you're likely to bump into. Some rocky intertidal zones contain *rock pools,* which are important sources of water when the tide recedes.

So who, or what, lives in the intertidal zone? Well that really depends on the region and variations in topography (for example, sand, rock, grass, mangroves, ice). We get into that more in Chapter 5. For now, let's look at the three regions of the intertidal zone.

High intertidal

If you're standing on the coast and feeling as though someone left you high and dry, you're in the high intertidal region. Here, creatures are adapted to eke out a more terrestrial existence. Because the high intertidal region doesn't get a lot of wave action, its inhabitants need to be better equipped to survive heat, sun, lack of moisture, and saltier water. (Water becomes saltier as it evaporates, leaving the salt behind.)

Common critters found higher in the intertidal zone are invertebrates that are better adapted to resist *desiccation* (drying out), such as snails, limpets, barnacles, and other animals with shells. Some of these animals store seawater in their shells to prevent them from drying out, sort of like carrying around their own reusable water bottles.

Middle intertidal

One step down from the high intertidal region is the middle intertidal region — home to anemones, chitons, and mussels, which feed underwater, yet have some degree of evolutionary adaptation to survive hot and dry spells. Sea stars and crabs may also wander up to this region for a visit.

Low intertidal

Life in the low intertidal region is more aquatic; no surprise there, because everything's submerged most of the day. Here, you find soft-bodied organisms, such as sea stars, sea slugs, sea urchins, and sponges, maybe even the occasional octopus (Philippe's favorite sea creature), as well as *turfing* algae and seaweeds (marine plants that form mats).

Algae is a staple of intertidal communities, forming the basis of most food webs.

COOL AND CREEPY

FUN FACT

You wouldn't know it from looking at them, but sea stars are vicious predators — carnivores that can muscle open mussels and clams using their five mighty suction-cup-equipped arms. But that's not even the coolest part. After gaining entrance, they literally spill their guts, ejecting their stomachs from inside their bodies, through their mouths to digest their prey. After dining, they suck their stomachs back in (not just in a vain attempt to look more fit). And they wonder why they can't get anyone to come over for dinner.

Plants and animals living in the low intertidal region aren't as well-adapted to hot spells and dry spells as organisms that live higher up in the intertidal zone. The good news is that they're generally well-adapted to getting body slammed by waves that would otherwise push or pull them out to sea. If you've ever been rocked by a big wave, you know the feeling . . . well, actually you don't really, because you're a lot bigger than a tiny little crab. Inhabitants of the intertidal zone must feel like the world is crashing down on them every few seconds. But they're tough, they've adapted, and now, for them, it's just the way life is.

Wading through the neritic zone

When you start to need some sort of flotation device (or a boat) to keep your head above water, you're in the *neritic zone* (also known as the *sublittoral zone*). This area starts at the low tide mark and continues to the edge of the continental shelf in the ocean — about 200 meters (660 feet) deep. While that may sound really deep, it's relatively shallow compared to how deep it gets out in the open ocean, which is much, much, much deeper.

Biodiversity is at its peak in the neritic zone, and no wonder — everyone wants to live here! Virtually every marine plant and thousands of coastal animal species (vertebrates and invertebrates) call this zone home. Consequently, the neritic zone is where 90 percent of all marine life resides. Diverse ecosystems flourish in and around coral reefs, seagrass beds, and kelp forests, all of which rely on sunlight-powered photosynthesis.

Why is it such a popular place? Well, conditions in the neritic zone are ideal for supporting diverse marine communities — the presence of sunlight throughout the zone; moderate, stable temperatures, pressures, and salinity; plenty of oxygen and carbon dioxide; and an abundance of nutrients that make their way from the land into the ocean (which under normal circumstances is good but can be bad when it's too much).

In fact, the neritic zone is such a pleasant place, it's also a popular hangout for sea creatures that move between deeper open ocean and shallow coastal waters, such as some sharks, turtles, and dolphins. And that's not all — common visitors include those that feed in coastal waters but can also haul out onto land, such as some marine reptiles, *pinnipeds* (seals, sea lions, walruses), and penguins.

FUN FACT

Even though the neritic zone accounts for only about 10 percent of the ocean, it produces roughly 90 percent of all the fish and shellfish we harvest, making it very valuable for us humans too.

Heading out to sea: The oceanic zone

The rest of the ocean that is neither intertidal nor neritic accounts for the oceanic zone. This vast, deep, inground pool comprises a *huge* chunk of Earth. It actually accounts for 95 percent of all the living space on the planet. Yet, it's the part of the ocean (and part of the planet) we know the least about.

It starts from the edge of the continental shelf (the outer edge of the neritic zone) and extends all the way down to the seafloor 11,064 meters (36,300 feet) at its deepest point at the bottom of the Mariana (or Marians) Trench. In fact, it's so deep, it had to be divided into zones, but let's not get too deep into that here; depth is the topic of the next section.

While the diversity of species in the oceanic zone can't hold a candle to the diversity in the neritic zone, the contrast in how life survives at the top and bottom of this zone can't be beat. Near the surface, where sunlight penetrates, plants anchor the food webs and ecosystems. In contrast, on the seafloor, where it's pitch black, ecosystems depend on either the remains of organisms that settle to the bottom or life that develops around hydrothermal vents and the bacteria that dine on the chemicals spewing from those vents.

So, what sort of life exists in the oceanic zone? Keep reading.

FUN FACT

HOW IS THE OCEAN LIKE A DESERT?

Despite the oceanic zone's enormity, it's actually considered to be a *marine desert* because it's so sparsely populated. When comparing the size of the oceanic zone to the number of species it contains, biodiversity here is relatively low.

Exploring the Five Vertical Zones of the Water Column

The ocean's *water column* (a conceptual pillar of water measured from the ocean's surface to the seafloor) is often divided into five zones — the epipelagic, mesopelagic, bathypelagic, abyssopelagic, and hadalpelagic zones (see Figure 4-2). The divisions generally correspond to differences in depth, amount of sunlight, temperature, pressure, nutrients, and organisms that live in those zones. In the following sections, we take a deep dive into each of the five vertical ocean zones.

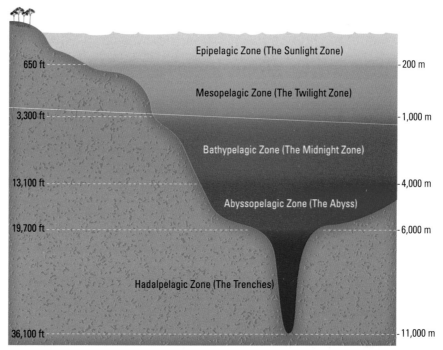

Epipelagic Zone (The Sunlight Zone)

650 ft — 200 m

Mesopelagic Zone (The Twilight Zone)

3,300 ft — 1,000 m

Bathypelagic Zone (The Midnight Zone)

13,100 ft — 4,000 m

Abyssopelagic Zone (The Abyss)

19,700 ft — 6,000 m

Hadalpelagic Zone (The Trenches)

36,100 ft — 11,000 m

FIGURE 4-2: The ocean's vertical zones.

©John Wiley & Sons, Inc.

Skimming the surface: The epipelagic zone

The *epipelagic zone* (commonly referred to as the *sunlight zone*) is the top 200 meters (about 650 feet) of the ocean, where enough sunlight is available for plant life to grow and support a large, diverse population of marine life. Because it forms the ocean's surface, the epipelagic zone experiences greater variations (compared to the other vertical zones) in temperature and other conditions due to climate, local weather patterns, and proximity to large land masses.

Who lives here? Lots of *plankton* (tiny plants and animals that float, as shown in Figure 4-3); *nekton* (tiny plants and animals that swim); jellyfish; sea turtles (see Figure 4-4); a variety of fish including tuna (see Figure 4-5), marlin, salmon, and sharks; and cetaceans (dolphins and whales).

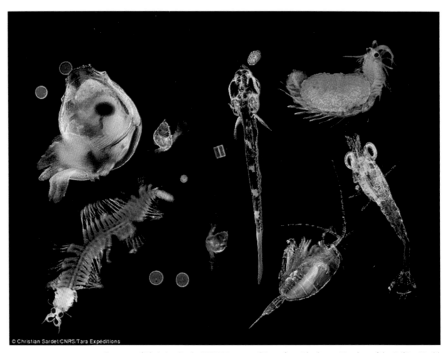

FIGURE 4-3:
Plankton.

Courtesy of Christian Sardet/CNRS/Tara expeditions, from Plankton—Wonders of the Drifting World, Univ Chicago Press 2015. Licensed under CC BY-SA 4.0

REMEMBER

It's not as though these zones are sectioned off like office cubicles. Plenty of animals roam freely from one zone to another. Animals that need to breathe air, such as sea turtles, dolphins, and whales, generally hang out closer to the surface just so they can get their heads (or noses or blowholes) above water regularly but may dive deep into the other zones to find food. Some deep-sea creatures may do the opposite and venture up from the deep to shallow water for other benefits, such as food, light, warmth, and lower water pressure.

Dimming the lights in the mesopelagic zone

Just below the epipelagic zone is the *mesopelagic zone* (commonly referred to as the *twilight zone*). This is the region spanning 200 to 1,000 meters (650 to 3,300 feet) below the surface, where some sunlight still penetrates, but not enough for photosynthesis. To feed, most animals in this zone move toward the surface. Others eat whatever *detritus* (table scraps) and organic matter fall from the epipelagic zone, or they just eat their smaller or weaker twilight zone neighbors.

FIGURE 4-4: Sea turtle.

Source: Cristina Mittermeier – www.sealegacy.org

FIGURE 4-5: Tuna.

Source: Keith Ellenbogen – www.keithellenbogen.com

Some animals in this zone have evolved the ability to produce their own light — a trait referred to as *bioluminescence* (creating light through biochemical processes). Instead of carrying a flashlight to find their way in the dark, they *are* the flashlight. Although scientists aren't quite sure about the purpose of this superpower, they think it might be used to ward off or evade predators (see the nearby sidebar on counter-illumination), detect or lure prey (ooh, shiny light!), or communicate with members of their own species.

Life starts to get a bit weird in this zone. Here, you're likely to start bumping into cool fish like the lanternfish, hatchetfish, and barbeled dragonfish, all of which can produce their own light. You can also find species of bristlemouths (it's okay, I brought my own toothbrush). These fish, generally no larger than your finger, are not only thought to be the most common fish in the ocean, but also the most common vertebrate on Earth — more abundant than humans, chickens, and rats combined. Let that sink in for a second. Some marine mammals and sharks can also be found here, but most will stay in the mesopelagic only for relatively short periods before returning to the surface. Swordfish (see Figure 4-6), *ctenophores* (see Figure 4-7) and *siphonophores* (jellyfish relatives), and firefly squid are other interesting animals that can be found in this zone.

Taking a deeper, darker dive into the bathypelagic zone

Just below the mesopelagic zone is the *bathypelagic zone* (also called the *midnight zone*), which extends from 1,000 meters to 4,000 meters (3,300 to 13,000+ feet) below sea level. No sunlight penetrates this zone, and the temperature is relatively constant at a very chilly 4 degrees Celsius (39 degrees Fahrenheit). Animals in this zone prey on other bathypelagic organisms or grab whatever organic matter rains down like manna from above. Some creatures in this zone migrate closer to the surface to feed at certain times of day.

FUN FACT

COUNTER-ILLUMINATION

Some bioluminescent sea creatures may use this skill as camouflage, illuminating their soft underbellies to blend in with light coming from the surface, while the tops of their bodies remain dark to blend in with the darkness below them. This application of bioluminescence, called *counter-illumination*, protects the creature from predators above *and* below. When predators from below look up, all they see is light. When predators from above look down, all they see is darkness. Take that, camo pants.

FIGURE 4-6:
Swordfish.

Joe Fish Flynn/Shutterstock

FIGURE 4-7:
Ctenophores.

Source: Schmidt Ocean Institute – www.schmidtocean.org

The creatures that live here are too insane to make up, but they're not the most colorful — just about everything is black or red, which makes them virtually invisible in water at these depths. (Certain wavelengths are filtered by water faster than others. Because red light has the longest wavelength and is absorbed quickest, once you go deep enough, anything red appears black.)

If you cut yourself diving at around 18 meters (60 feet) deep, your red blood may appear purple and, if you go any deeper, even black. Of course, we're not recommending that you poke your finger when you're diving, but if you happen to suffer a small cut underwater at that depth and you're looking for a cheap thrill . . . .

Calling this zone their home are the weird and wonderful barreleye fish, giant isopods, viperfish, vampire squid, and anglerfish. Occasionally you can find sperm whales here, and if you're really, really lucky, you can see one battling a giant squid (of course if you do, take a picture because no one has captured that epic battle on film yet). The deepest diving marine mammal, the Cuvier's beaked whale, can also reach this zone. This elusive and strange-looking animal holds the record for the longest mammalian dive, plunging up to 3,500 meters (11,480 feet) deep (that'll make your ears pop) in search of deep-water cephalopods and squid.

Many animals in this zone and deeper have adaptations to allow them to eat almost anything, including prey much larger than them. Gulper eels have specialized jaw structures that enable them to open their mouths incredibly wide (see Figure 4-8). Sharks and their relatives, including the Greenland shark (which can live for 400 years), ghost shark, frilled shark, and goblin shark can sometimes also be found in this zone, as well as the deepest living octopus, the dumbo octopus shown in Figure 4-9 (although some say it can be found at even greater depths).

Delving into the abyss: The abyssopelagic zone

One step down from the *bathypelagic* is the *abyssopelagic zone* (also called the *abyss*), extending from 4,000 to 6,000 meters (13,000+ to nearly 20,000 feet) below the surface. Imagine totally dark, near-freezing temperatures (though stable), and super high pressure. For animals adapted to these harsh conditions, the pressure is no problem. Unlike animals with gas-filled organs (such as lungs and swim bladders) that would be crushed at these depths, deep-sea creatures are pretty much made up of tissue and fluid. While the high pressure may limit species diversification, it isn't the hardest part about living here. The more challenging factor is the scarcity of food.

FIGURE 4-8:
The gulper eel;
say ahh!

Source: Woods Hole Oceanographic Institution, P. Caiger – www.whoi.edu

FIGURE 4-9:
Dumbo
Octopus; isn't
she (or he)
cute?

Source: Schmidt Ocean Institute – www.schmidtocean.org

Generally, the farther down you go, the fewer species you encounter, because these are tough environmental conditions to adapt to. Life here is thought to have changed little over millions of years. Some abyssal species include the common fangtooth, the tripod fish shown in Figure 4-10 (they're *hermaphroditic*, meaning they have both male and female reproductive organs, which means they can produce young either with another fish or on their own!), hagfish, cusk eels, grenadiers, and viperfish (Figure 4-11). In some places, you can find deep-sea corals, which don't need sunlight to survive.

FIGURE 4-10:
The tripod fish.

Source: Schmidt Ocean Institute – www.schmidtocean.org

REMEMBER

Deep-sea creatures must be able to tolerate intense pressure (from the weight of the water above), total darkness, and near freezing temperatures. (See Chapter 5 for more about these ecosystems that thrive under extreme conditions.) That's not to say that animals living closer to the surface have an easy life; they face a greater risk from predators and from changes in environmental conditions.

How low can you go? The hadalpelagic zone

The deepest zone in the ocean is the *hadalpelagic zone* (also called the *trenches*), which is anything deeper than 6,000 meters (about 20,000 feet) below the surface, such as in the deep ocean trenches. This realm is named after Hades, the Greek god of the underworld. We don't know much about this zone, because it's hard to get to and requires super specialized technology to cope with the immense pressure.

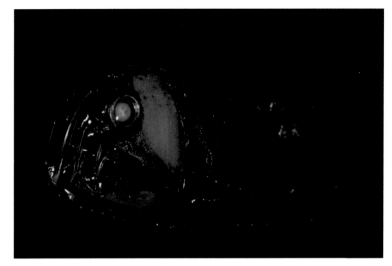

Source: Woods Hole Oceanographic Institution, P. Caiger – www.whoi.edu

Life becomes very limited in this zone. You can find giant, single-celled xenophyo-phores, deep-sea microbial mats, amphipods, sea cucumbers (see Figure 4-12), jellyfish, and other invertebrates such as tube worms (see Figure 4-13), decapods, bivalves, and sea-anemones. Species of snailfish, cusk eels, and eelpouts can also be found in this region but are limited to relatively shallow areas, and usually closer to the seafloor. And even though it is really hard for people to venture this far down, our trash manages to find a way — namely, our plastic. Yup, even here in the deepest part of the ocean, recent expeditions found a plastic bag at one of the deepest points of the Mariana Trench, nearly 11 kilometers (about 7 miles) down. Ugh.

FIGURE 4-12:
A sea
cucumber.

Source: Schmidt Ocean Institute – www.schmidtocean.org

FIGURE 4-13:
Tube worms.

Source: Woods Hole Oceanographic Institution, P. Caiger – www.whoi.edu

Acknowledging the Existence of Other Zones

Oceanographers have come up with other terms and concepts for understanding ocean zones that may be more useful in certain contexts. These other zones don't contradict the horizontal or vertical zones we present in the previous sections. They just provide a more general breakdown that may be easier to remember and use in dinner conversations. Knowing a little about these zones is also useful in case you encounter the terminology in the context of other discussions.

From light to dark: The photic and aphotic zones

The *photic zone* is basically the same as the *epipelagic zone* — the topmost 200-meter (650-foot) layer of the ocean. Personally, we prefer the term photic over epipelagic, because it's more descriptive (*photic* means related to light), easier to remember, and easier to pronounce. As with the epipelagic zone, the photic zone describes the layer of the ocean that has enough light for plants to grow.

Aphotic is the opposite of photic. It is the absence of light or, more precisely, insufficient light for plants to grow, which describes about 90 percent of ocean water from 200 meters deep to the seafloor. In the aphotic zone, animals can't rely on plants to anchor the food web. If they want to eat, they have three options:

>> Eat the table scraps that drop down from the photic zone.

>> Eat their neighbors (prey tell!)

>> Move to vent systems on the bottom (these are few and far between) where chemosynthetic bacteria (instead of plants) called chemoautotrophs anchor the food web. (*Chemoautotrophs* get their energy from oxidizing inorganic compounds instead of capturing energy from the sun through photosynthesis. *Photoautotrophs* capture energy from the sun through photosynthesis.)

FUN FACT

Some deep-sea organisms establish symbiotic relationships with chemoautotrophs. For example, giant tube worms, which hang out near deep-sea hydrothermal vents, provide a place for the bacteria to live in exchange for organic matter (food). Without the bacteria, the worms couldn't survive, and without the worms, the bacteria wouldn't have the "comfortable home" they need to grow and prosper.

From top to bottom: The pelagic and benthic zone

One of the simplest ways to section off the ocean is to divide it into top and bottom — the top being the *pelagic zone* (nearly all the water), and the bottom being the *benthic zone* (consisting of a relatively thin layer of water above the seafloor, the seafloor itself, sediment, and some subsurface layers).

Organisms may be classified as either pelagic or benthic. Pelagic organisms inhabit the open sea and surface of the ocean, whereas benthic organisms hug the bottom. Species like corals, seagrasses, mollusks (oysters, mussels and the like), crustaceans (crabs, barnacles, and so forth), and fish like rays and flounder are benthic; basically anything that's attached to the bottom or spends most of its life crawling or swimming along the bottom is considered benthic. All other organisms that float or swim are pelagic.

Chapter **5**

Checking Out the Neighborhoods: The Ocean's Ecosystems

An *ecosystem* is a biological community of organisms interacting with their physical environment as a whole. On land are terrestrial ecosystems — grasslands, deserts, tundra, and various types of forests. In and around the ocean are marine ecosystems — marshes, tidal zones, estuaries, mangrove forests, coral reefs, and the list goes on.

Think of ecosystems as distinctive countries around the world or neighborhoods within a country, each with its own terrain, architecture, residents, economy, and customs. While each ecosystem more or less functions as a self-contained unit they can overlap with neighboring ecosystems to varying degrees.

In this chapter, we take you on a tour of the ocean's varied ecosystems and some of the animals that call them home and then introduce you to a few of the marine creatures that travel around the ocean as if it were one continuous ecosystem (which it is, but shhhhh, that will be our little secret).

Hugging the Shore

Shoreline (littoral) ecosystems are those that follow the coastlines, where land and water meet. Think of them as the transition from land to sea. They're generally wet or at least damp most of the time, and salty, but water and salt content vary based on how far inland you go and on the terrain and climate. In some areas, you may find an entire ecosystem living in and around a body of water not much larger than a puddle, whereas other shoreline ecosystems are beneath the ocean's surface most, if not all, the time.

In this section, we lead you on a meandering journey down the shoreline to visit a variety of ecosystems and meet the characteristic and fascinating residents of each.

Digging life in the sand

What a great feeling it is to spread your beach blanket out on the sand and relax the afternoon away in peace and solitude. Well, not to ruin it for you, but you're not alone. Even though you may not see any people milling about, you're actually lying on a very crowded beach teeming with life, and we're not just talking about those seagulls trying to steal your lunch. Above and below the surface of the sand are a variety of life-forms, most of which are too tiny for you to see.

But don't let their tinyness fool you; among these life-forms are some really tough cookies. After all, beach living is a hard-knock life. Imagine being small and living on a beach. You go from being underwater, wet and salty one moment, to being exposed, dry, and hot the next. If you're small enough, you have to deal with the tight squeeze between sand particles, and if you're large enough, those same particles are constantly scraping against you and getting into your eyes and nose and, well, everywhere else. To top it all off, unless you're lucky enough to live in a bay or other protected area, you're tossed around all day by unrelenting waves.

Living here are tiny microbes too small to see, along with algae, plankton, sand dollars, crabs, and snails. Big animals like the sandy shoreline, too, including fish, rays, and sharks, which scour the shallow waters for food; turtles (for nesting); birds (for feeding); and pinnipeds (for napping).

One notable inhabitant that may have you grabbing your blanket and heading back to the motel is the innkeeper worm (see Figure 5-1). They generally hang out below ground (thank god for small favors) but occasionally wash up on beaches by the thousands when they get spooked out of their burrows.

FIGURE 5-1: The innkeeper worm.

Photo by Jerry Kirkhart. Licensed under CC BY 2.0

Living life on the rocks

To live on a rocky seashore, you have three options: Find a tide pool to hang out in, wedge yourself in a nook or cranny, or hang on for dear life. (A *tide pool* is a puddle in a rocky depression that fills with sea water at high tide, remains relatively full at low tide, and is refilled at high tide with fresh sea water and nutrients.)

The tide pool option is the cushiest. It's sort of like having your own apartment with a bunch of diverse and colorful roommates — mussels, snails, starfish, barnacles, anemones, urchins, limpets, crustaceans, seaweed, and even small fish and octopus. Nooks and crannies are a little more exposed to waves and wind, so they tend to favor animals that can either hold on or retreat deeper into holes when the going gets rough, such as crabs, sea snails, chitons (pronounced "kite-ins"), and sometimes starfish. The harshest real estate on the rocky shore is right where the water meets the land. All these residents can do is hold on for dear life as they are frequently battered along these rocky high-energy coastlines. (*High-energy* refers to frequent large waves crashing against the shore.)

WHERE DID ALL THIS SAND COME FROM?

Sand is a collection of crushed, rolled and cracked quartz (light colored sand), calcium carbonate (white sand), volcanic glass (black sand), and other substances depending on the location. Sand can also come in pink, red, and green hues due to algae or other mineral deposits in the area.

White sand consists of crushed up pieces of shells and skeletons of reef-living marine organisms, including corals, mollusks, and microorganisms that are usually ground into fine grains by the power of waves and wind. However, a certain percentage of white sand is also poop, specifically parrotfish poop! Parrotfish (see Figure 5-2) feed by scraping algae off coral with their hilarious giant teeth (shaped more like a beak). Some coral gets chewed off as well and comes out the other end, in the form of tiny chewed up pieces of calcium carbonate. On average, one green humphead parrotfish can produce 90 kilograms (200 pounds) of sand each year. So yes, parrotfish — poop — sand. Think about that during your next romantic interlude on a beach!

FIGURE 5-2:
Terminal bullethead parrotfish headed in for a bite of coral.

Photo by Kevin Lino NOAA/NMFS/PIFSC/ESD Licensed under CC BY 2.0

Two of the toughest hangers-on are barnacles and chitons. Barnacles are *sessile* (immobile) crustaceans that commonly cement themselves to rocks, piers, bridges, and boat hulls. Imagine stationary crabs in cone-shaped shells they grow themselves (see Figure 5-3).

Chitons are mollusks that look like turtles without heads, legs, or tails (see Figure 5-4). They can lie flat to wedge themselves between rocks or curl up like armadillos. You can tell a chiton when you see one by the eight overlapping armored plates on its back.

REMEMBER

Rocky shores are also home to larger animals that spend a significant amount of time out of the water, including penguins, a wide variety of sea birds, and pinnipeds like seals, sea lions, and elephant seals.

Mixing it up in the estuaries

Estuaries are partially enclosed areas along the coastline where fresh water from rivers and streams flows into the ocean. Here, fresh water and salt water mix, causing *brackish* water (not too salty, not too fresh, but somewhere in the middle). The salinity of the water changes throughout the day and the seasons depending on the tide, air temperature, rainfall, and other conditions.

If that's not enough to mix you up, depending on the estuary, it can include a variety of habitats such as coral reefs, oyster reefs, rocky shores, mudflats, salt marshes, and mangroves. For example, Chesapeake Bay has three different habitats — oyster reefs: home to oysters (duh!), mud crabs, and small fish; seagrass, where seahorses, blue crabs, and fish hang out; and open water, where you can find sea turtles, rays, and more fish.

Estuaries are also home to the largest reptile in the world — the estuarine croco-dile (shown in Figure 5-5), which can grow up to 6 meters (21 feet) long. Crikey! That's one *huge* reptile with a nasty temper, to boot, but don't worry — they only live in the tropical estuaries of Australia and Southeast Asia.

REMEMBER

Estuaries are in constant decline due to coastal development, dredging, overfish-ing, and pollution.

Muddling through the mudflats

A *mudflat* (a.k.a. *tidal flat*) is just what it sounds like — a flat, barren expanse cov-ered in mud that usually forms where river meets sea. The river carries particles of mud downstream and dumps them in the ocean. When the tide comes in, it slows the flow of the river allowing the particles to settle. Typically, wave action in these areas is minimal, so the fine particles aren't washed out to sea.

Mudflats are covered with water at high tide and uncovered at low tide, at which time they can become some of the muckiest, most foul-smelling places on the planet, at least to humans. The stench comes from the swamp gases, methane (CH_4), and hydrogen sulfide (H_2S), produced by bacteria that live in the mud and help break down organic matter to make it more digestible. These bacteria, along with algae living on the surface and *detritus* (decomposing plants, animals, and other organic matter), provide a rich source of food to anchor the food web. Protozoa and nematode worms feed on the bacteria; crustaceans, mollusks, and polychaete (pronounced "poly-keet") worms feed on the bacteria, detritus, and the nematodes; and larger animals in turn feed on them. Mudflats are also vitally important stopover areas for migrating shorebirds who rely on mudflat residents such as crabs, fish, and mollusks to fuel their long journeys.

FIGURE 5-5: An estuarine crocodile (a.k.a. saltwater crocodile).

Pius Rino Pungkiawan/Shutterstock

REMEMBER Looking at a mudflat, you might conclude that it's a smelly wasteland, and you wouldn't be the first. People have destroyed many of these valuable ecosystems by filling them in and building on top of them in the name of coastal development. But that's a big mistake because, in addition to serving as a huge source of food to coastal ecosystems, mudflats also help prevent coastal erosion. They're also wonderful places to hike and watch birds. However, if you decide to hike a mudflat (an activity cleverly called "mudflat hiking"), be sure you go with a guide who knows the tides well. Muck, especially sandy muck, is like quicksand. If you don't know what you're doing, you may find yourself up to your chest in muck with the high tide rushing in.

Settling down in salt marshes

Salt marshes are like mudflats but *with* vegetation. They usually form after enough mud and *peat* (decomposing plant matter) has accumulated to raise the ground level to a point at which it's high enough to support the growth of saltwater grasses and other plants that have adapted to living in shallow salt water. Salt marshes are flooded and drained daily or at least occasionally by the overflow of ocean water, which ensures a steady supply of salt.

Like mudflats, salt marshes are home to snails, mussels, worms, and crabs, but they also attract fish and shrimp looking for food or a place to lay their eggs amidst the relative safety of the plants that offer plenty of nooks and crannies to hide in. Ducks, geese, wading birds, and many migratory bird species also come to feed on the abundant vegetation, seafood, and insects. Mammals are attracted to salt marshes, too, with mice, rats, and raccoons coming to feed on the bounty; even the occasional coyote has been known to frequent salt marshes for the fine dining experience.

Clearly, salt marshes are very important to an enormous range of species . . . including humans. They provide essential food, refuge, and/or nursery habitat for various fisheries species that are vital to many coastal economies, including shrimp, blue crab, and numerous fish. In addition, they filter runoff water and help fight erosion.

FUN FACT

Salt marshes can be found all over the world. In the United States, half of all the salt marshes are located along the Gulf Coast of Texas, Louisiana, Alabama, Mississippi, and Florida, but thanks to oil and gas development they are disappearing fast.

Meandering through the mangroves

Mangroves are a group of tropical and subtropical trees and shrubs that live in the intertidal zone. They range in size from small shrubs to huge trees that are 12 meters (40 feet) tall and are adapted to living in salt water. Some species get rid of excess salt through their leaves while others have special pores that keep much of the salt from entering in the first place.

So, mangrove trees are cool, but what have they done for us lately? Well . . . a lot, including the following:

>> Provide a habitat for algae, insects, mollusks, crustaceans, fish, sharks, birds, dolphins, manatees, turtles, lizards, crocodiles, alligators, snakes, frogs, small mammals, a few marine mammals, and more insects than you can probably imagine.

>> Serve as a nursery where many fish and other sea creatures lay eggs, give birth, and protect and feed their young. Mangrove forests are directly or indirectly responsible for a considerable percentage of the global fish catch.

>> Protect the coastlines and coastal communities from storm surges, hurricanes, typhoons, waves, and floods. (Every 101 meters [330 feet] of mangrove forest can reduce wave height by up to 66 percent.)

>> Filter the water, improving clarity and quality.

>> Store carbon, which mitigates global warming. One acre of mangrove alone can sequester 1,450 pounds of carbon a year, which is the amount of carbon produced by driving your car across the United States three times. That's also more carbon than can be stored by an acre of any inland forest. (See Chapter 21 for details.)

>> Contribute to the world economy. Worldwide, mangroves contribute about $1.6 billion each year to local economies.

Mangroves not only protect land but also create it. Mud collects around the network of roots, leading to the development of shallow mudflats.

Any discussion around the mangrove ecosystem is more like a tale of two ecosystems — the one below water and the one above it (see Figure 5-6) — as explained next.

FIGURE 5-6:
This small
mangrove
cluster off the
coast of Saudi
Arabia already
supports life
above and
below the
surface.

Source: Shane Reynolds – www.ColorEarth.tv

Living among the roots

Some mangroves have roots that reach from above the surface of the ocean down into the substrate below. This enables them to survive the ups and downs of the tides, but it also creates a tangle of roots that makes an ideal habitat for small ocean creatures, including the following:

» *Invertebrates* (animals without backbones): Snails, barnacles, bryozoans, tunicates, mollusks, sponges, polychaete worms, isopods, amphipods, shrimp, crabs, and jellyfish all live in or near the network of roots.

» **Fish:** A wide variety of fish live inside or adjacent to the root system, many of them, such as sharks, in the juvenile phases of life where they're safe from bigger fish that can't get in between the roots. Fish types vary according to water temperature, salinity, turbidity (cloudiness), and other characteristics of the water. In addition, salinity can change with seasons, drawing more or fewer freshwater species that can tolerate some salinity.

» **Marine mammals:** Farther out from shore, you can find some marine mammals, including bottlenose dolphins, manatees, and dugongs. Dolphins are typically in search of fish, whereas manatees dine on the vegetation.

Keeping your head above water

Above the waterline, mangroves support an even greater diversity of creatures, most of which are terrestrial or amphibious, including the following:

>> **Reptiles:** Crocodiles, alligators, various snakes and lizards, and land turtles and (farther offshore) sea turtles. Also, some freshwater turtles nearer to the headwaters may frequent the mangroves.

>> **Amphibians:** Amphibians aren't generally drawn to mangroves because of the salinity of the water, but you may find some tree frogs and toads.

>> **Birds:** Many species of birds live in the mangroves or feed there, including egrets, heron, spoonbills, ducks, grebes, loons, cormorants, falcons, eagles, owls, vultures, and more.

>> **Mammals:** Mammals that frequent the mangrove forests include panthers, raccoons, skunks, mink, river otters, bobcats, possum, rabbits, rats, and even deer.

Swimming through Kelp Forests

Imagine flying free through a giant forest surrounded by beautiful colors and bizarre, extraordinary animals (see Figure 5-7). That's what it's like to swim through a kelp forest. While we may call them kelp forests, kelp is not technically a plant like a tulip or a tree. In fact, *kelp* is a multicellular algae that just looks like a plant. What distinguishes it from a plant is that plants have a vascular system to transport water and nutrients, whereas every cell in kelp is responsible for absorbing its own water and nutrients.

FIGURE 5-7: Shafts of light stream down through the canopy of a kelp forest off the coast of California.

Source: Hal Wells – www.HollywoodDivers.com

Smaller gatherings of kelp are called *kelp beds.*

Kelp thrives in cool/cold (from 42 to 72 degrees Fahrenheit), nutrient rich, and relatively shallow 2 to 30 meters (6 to 100 feet deep) water. It attaches itself to the seafloor with small rootlike structures called *haptera* or *holdfasts.* The kelp stays vertical in the water column thanks to gas filled floats (like tiny buoys) at the top of their *stipes* (comparable to stalks) or where each blade (the leafy part) branches off. Some kelp can grow up to 46 centimeters (18 inches) per day and reach up to almost 50 meters (150 feet) long.

These underwater forests are high in biodiversity. Like terrestrial forests, they can be divided into three layers, which are a little different for kelp forests —the *canopy* lying on the surface, the *subcanopy* or *understory* below the surface, and the *forest floor* where the kelp is anchored. The canopy is home to small crustaceans, snails, sea slugs, larvae of many types of marine animals, and juvenile fish of all sorts. You're likely to see sea otters, seals, sea lions, and a variety of birds that frequent the area to feed. The subcanopy is home to fish, sharks, squid, and, in Southern Ocean kelp forests, one of my (Philippe's) favorite creatures, the weedy sea dragon (see Figure 5-8). On the forest floor, you're more likely to see snails, starfish, oysters, lobsters, crabs, and my (Ashlan's) bestie, various species of octopus.

FIGURE 5-8
Weedy sea dragon off the southern coast of Australia.

Source: Sheree Marris – www.shereemarris.com

As quickly as kelp forests grow, they can be destroyed. Global warming is a huge problem. Other threats include pollution, overfishing, invasive kelp species, and voracious sea urchins. A group of sea urchins can destroy entire kelp forests at a rate of 9 meters (30 feet) per month. Fortunately, urchins are a delicacy to sea otters, which eat 20 to 30 percent their body weight in food every day. Unfortunately, sea otters are facing their own challenges — from not enough fish to eat and increased pollution to human related disease and even poaching.

The absolute cutest activity that takes place in kelp forests (and maybe in the entire ocean) is this: Sea otters wrap their babies in kelp so they don't drift away while the mother hunts or takes a well-deserved See Figure 5-9 for a photo of a mother snuggling with her baby.

FIGURE 5-9:
Too cute —
mother and
baby sea otter.

Source: Lootas and pup. C.J. Casson, Seattle Aquarium

Swirling in Sargasso: A Sea without Borders

The Sargasso Sea is an area in the Atlantic Ocean where vast mats of a unique brownish/gold seaweed called *sargassum* live. Like kelp, sargassum has gas-filled bladders that make it float. Unlike kelp and other seaweeds, it's *holopelagic*, meaning it spends its whole life ("holo") floating in the *pelagic zone* (the water column) without needing contact with any land form and reproducing *vegetatively* (new plants grow from fragments of parent plants). Pretty cool, huh? The Sargasso Sea is also the only sea without a land boundary.

Sargassum gathers in and around the North Atlantic Gyre, a huge, swirling body of water off the eastern coast of the United States that spins due to ocean currents (see Chapter 16). These swirling currents concentrate the sargassum into enormous mats (called *windrows*), which can be hundreds of meters long. Based on satellite images, scientists estimate the biomass in the Sargasso Sea to be around one million tons.

The Sargasso Sea is an oasis in the open ocean for all types of creatures, including sea turtles, whales, sharks, fish, crabs, snails, octopus, squid, worms, shrimp, nudibranchs, and more. Some animals are specially adapted to live in this floating haven, such as the iconic sargassum anglerfish, which not only blends into the surroundings but also has special fins that enable it to climb through the seaweed (see Figure 5-10).

FIGURE 5-10:
The sargassum anglerfish.

However, while some creatures spend their whole life in the sargassum, others just take a break while passing through. The most famous of these are sea turtles. After emerging from their nests, sea turtles scurry down the beach into the ocean. In the Atlantic, many species of sea turtles then swim as fast as they can out into the Sargasso Sea to spend their early years hiding amongst the relative safety of the sargassum mats before they are big enough to venture out again into the open ocean.

Grazing in the Seagrass Meadows

Unlike kelp and sargassum (both seaweeds) seagrass is a bona fide plant, complete with a *vascular system* (a network of veins to move nutrients and dissolved gases throughout the plant). In fact, it's the only *angiosperm* (flowering plant) in the ocean. The flowers are usually borne underwater, and pollination occurs underwater as well (in most cases). Seagrass also has a true root system, as opposed to a simple holdfast like kelp. The root system anchors the plant to its substrate and acquires nutrients for the plant as well. Similar to lawn grass, seagrass roots commonly send out *rhizomes* (underground shoots that sprout more grass). This root/rhizome system creates a dense, interwoven mat that holds the grass and seafloor in place, thereby literally anchoring the entire ecosystem.

Seagrass can be found nearly everywhere in the world except Antarctica. It grows in both salty and brackish water, as long as the water is shallow, and can vary in size from small mats to enormous meadows that resemble fields of underwater wheat. Botanists estimate that globally these seagrass fields produce more vegetation than all the wheat fields of the U.S. grain belt combined. Some of these fields are even large enough to be seen from space!

Seagrass fields are extremely important ecosystems. They provid nursery and adult habitat and feeding grounds for many species, from small invertebrates all the way up to turtles, birds, crabs, and large fish. Most seagrass grazers use the broken up and degrading leaves as their food source. Others, such as the sea turtle, the manatee (see Figure 5-11), and its relative the dugong, eat the entire plant — and a lot of it too. One large, hungry manatee can gulp down almost 100 pounds of seagrass a day.

In addition to providing food and shelter, the seagrass helps reduce erosion, filter water, release oxygen into the water, and extract and sequester carbon dioxide. With all these benefits, seagrasses are thought to be one of the most valuable ecosystems in the world, just behind coral reefs, estuaries, and wetlands. One hectare of seagrass (about two and a half acres) is estimated to be contribute over $19,000 per year of value to the global economy.

FIGURE 5-11:
Manatee
and calf.

Photo by Sam Farkas, NOAA OAR Photo Contest 2014. Licensed under CC BY 2.0

Building Their Own Communities: Reefs

A *reef* is a ridge of rock, sand, or coral near the surface of the water, but because this chapter is about ecosystems, we're going to focus on reefs that are teeming with life — coral reefs and oyster reefs. Coral and oysters, respectively, are responsible for building the foundation on which these living reefs grow.

Coral reefs

While coral reefs may seem like lifeless rocks, they are actually incredibly complex ecosystems created by a marine invertebrate called a *coral polyp*. Classified as *cnidarians* (pronounced ny-*dare*-ee-uhns), they're related to jellyfish and sea anemones (see Chapter 14). Over 2,500 species of coral exist, and they can be broken down into two groups: hard (rigid) and soft (flexible). The hard ones (such as brain coral and elkhorn coral) are referred to as "the reef builders" because they create the hard material that forms the basis for the ecosystem. The soft ones (such as sea fingers and sea whips) hang out in and around reefs. Coral reefs exist in relatively shallow water (usually no deeper than around 30 meters [100 feet]) because the algae (zooxanthellae) that live symbiotically inside the coral providing them with essential nutrients, require sunlight.

Coral polyps (see Figure 5-12) build large reefs through the process of creating small homes for themselves out of calcium carbonate that are tightly packed next to other coral polyps that do the same thing. Reefs grow when subsequent generations of coral polyps stack their homes on top of their deceased predecessor adding height and volume to the structure. As they create more and more homes, they end up with a massive system similar to a human apartment complex which is formally known as a "colony." All the polyps within a colony are genetically identical, and the colony as a whole is considered a single living organism.

As more and more colonies are built (by the same or different species of hard coral), a coral reef is formed. Depending on its size, a healthy reef could have hundreds of thousands to billions of diverse coral polyps living together in close quarters, just like a large human city. Different colonies even engage in turf wars with the polyps of one colony stinging those of the other colony. (Can't we all just get along?) So, what may look like a bunch of rocks, is actually a living ecosystem with a dazzling array of colors and shapes that would make Picasso cry.

FIGURE 5-12: Each oval structure houses an individual coral polyp.

Source: Cristina Mittermeier – www.sealegacy.org

But that's not all. In the process of building homes for themselves, coral polyps lay the foundation for rich ecosystems that provide food and shelter for highly diverse populations of other marine organisms, including thousands of species of fish, invertebrates, plants, algae, sea turtles, birds, and marine mammals. The largest coral reef on the planet is the Great Barrier Reef, and it can be seen from space. Coral reefs are some of the most stunning places on the planet (see Figure 5-13).

Soft corals also build their own homes but they do so out of spongy material, so the homes are more flexible and fragile. (We know, this is sounding like the story of the three little pigs, right?) While soft corals don't qualify as reef builders, they're a beautiful part of some reef systems.

Source: Shane Reynolds – www.ColorEarth.tv

FIGURE 5-13:
Coral reefs like this one off the coast of Saudi Arabia are rich in diversity.

REMEMBER

The value of coral reefs extends far beyond their beauty and the fact that they provide food and shelter for a rich community of marine life. They're also the first line of defense against waves and storms that threaten to batter the coasts. The World Wildlife Fund and the Smithsonian Institute estimate the commercial value of coral reefs worldwide at a minimum of $1 trillion per year, $300 to $400 billion of which is attributed to food, tourism, fisheries, and medicines.

FUN FACT

Coral is actually white because it's made from calcium carbonate, which is essentially limestone. In fact, most limestone is actually ancient coral reefs, and if you look closely at limestone buildings, sometimes you can see the outline of the coral polyps and ancient shells that lived on the reef. What gives living corals color are the *zooxanthellae* (algae) that live in a *symbiotic* (mutually beneficial) relationship with the coral polyp. The zooxanthellae, which live inside the coral polyps, capture energy from the sun and turn it into food (like a plant), which is shared with the coral polyps. In exchange, the coral provides the zooxanthellae with a place to live and chemical compounds needed for photosynthesis.

FUN FACT

Corals come in a variety of shapes, which vary mostly by the species that build the colonies, but their shape is also influenced by the wave action where they live. In more turbulent waters, they tend to grow as mounds or flattened shapes, like tables, whereas in calmer seas they form more intricate branching structures.

CORAL REEFS AT RISK

Coral reefs are disappearing at a terrifying rate. Up to 40 percent of the world's coral has died in the last few decades, including half of Australia's Great Barrier Reef. How could this happen? Well, coral may be hard as a rock, but the coral polyps that build it are sensitive to changes in their habitat.

Two main factors threaten the coral reefs — heat and pollution. First consider the impact of heat. As the temperature of the ocean rises (due to global warming), in many places it frequently passes the threshold tolerable for corals to live. When that happens — the zooxanthellae living inside the polyps begin to produce toxins, or they become parasitic, so the coral essentially kicks them out of the house. And without the zooxanthellae, the polyps starve. The second threat is pollution in the form of excess nutrients in runoff, chemicals in plastics and sunscreens, and other sources, which can kill the zooxanthellae, the coral, or both.

Because zooxanthellae are the source of the coral's color, when they're gone, the coral turns white, a phenomenon called *coral bleaching*. Even if the coral polyps are still alive, all you see is the white because polyps are basically transparent without their zooxanthellae. Unfortunately, even if the polyps survive, they begin to slowly starve to death. So, while coral bleaching isn't necessarily a death warrant, it's a serious warning sign that the coral is stressed and at increased risk of death. If water conditions don't return to normal relatively quickly, the polyps will die, and all that beauty, mystery, wealth and diversity dies with them. The following figure shows healthy coral on the left and bleached coral on the right. The photos are from the same sight before and after a surge of warm water.

Source: Underwater Earth / XL Catlin Seaview Survey – www.underwater.earth

Oyster reefs

Not to be outdone by the coral polyps, oysters create their own colonies and, in the process, create rich ecosystems that provide food and shelter for a thriving marine community, not as varied as coral mind you, but still pretty diverse.

So, if hard corals build coral reefs, how do oysters go about building oyster reefs? Well, it all starts with baby oysters (okay, technically oyster larvae that develop into juveniles are called *spat*), which, at this stage of their lifecycle are free-floating. At the mercy of currents and tides, they float around aimlessly in salty or brackish water until they find something to latch onto — a rock, a shell, even another oyster. After taking hold, they grow and release more larvae, some of which settle down nearby, and the cycle continues until a structure is formed worthy of being called a reef.

These reefs are wonderful places for other benthic marine animals to latch onto, including mussels, barnacles, and sea anemones. Collectively, all these animals and the creatures attracted to the structures they create form an abundant food source for many species of fish and other marine animals, not to mention humans who love slurping down oysters on the half-shell. Along with food, oyster reefs provide a nursery ground for many creatures, including shrimp, herring, striped bass, flounder, blue crabs, and more.

FUN FACT

Oysters and other bivalves (clams and mussels) are incredible little water filtration machines. A single oyster can filter 7.5 liters (2 gallons) of water every hour! Just search online for "oyster filtration video" to see them in action — it's mind blowing.

Oyster reefs are important for water quality, fish stocks, human food, and (like coral reefs) they protect the coastlines from waves, tides, and storms.

Chilling Out at the Poles

Polar marine ecosystems, formed at the North Pole and the South Pole, are critical to the entire ocean, even to other ecosystems located as far away as the equator. While they're too cold for us to enjoy a refreshing swim, the frigid waters of polar ecosystems are actually bursting with life. Massive amounts of nutrients exist at the poles and are transported around the world by powerful ocean currents (thanks to differences in water temperature and salinity, wind especially near the South Pole, the rotation of Earth, and other forces).

These currents are much more prevalent around Antarctica, where the Antarctic Circumpolar Current (ACC) originates. The ACC is the largest wind-driven current in the world, connecting the Atlantic, Pacific, and the Indian Oceans. The ACC plays a big role in the global ocean conveyer belt (see Chapter 16). As sea water freezes, water and salt are separated. The water freezes, and the salt is left behind, which makes the remaining water saltier and denser. (Cold water is also denser than warmer water.) This cold, salty water drops down, pushing warmer water out of the way and, in the process, creating a slow-moving current that carries nutrients and cooler water to different parts of the ocean around the world.

In terms of their seasons, extreme cold, and the interplay between freshwater ice and salty seawater, both polar ecosystems are similar. The big difference between the two is that Antarctica (see Figure 5-14) is a continent (land mass) covered in ice, whereas the Arctic (see Figure 5-15) is an ocean covered by ice. Another way to put it is that the Arctic is a sea surrounded by land, whereas the Antarctic is land surrounded by sea.

FIGURE 5-14:
Ashlan in Antarctica in front of Mount Erebus.

Source: Ashlan Cousteau

Down south, the Antarctic ecosystem is fueled more by upwellings of nutrient-rich water from the deep that feed phytoplankton and ice-algae at the surface. Then, Antarctic krill eat the phytoplankton and algae, and larger animals, including fish, penguins, whales, and seals, feed on the krill and smaller fish. Some fish, such as the Arctic cod and Antarctic ice fish, contain proteins that function like antifreeze to keep them from freezing in sub-zero waters.

The Arctic ecosystem is highly seasonal. During the six months of summer sunlight, algae and phytoplankton grow like mad, producing enough food to support the ecosystem through the six-month winter. Zooplankton eat the algae and phytoplankton, larger predators eat the zooplankton, and so on, all the way up to walrus, narwhals, and whales such as beluga and minke. Meanwhile, bits of food discarded by these surface creatures drop down from above to feed the fish, sea stars, soft corals, and other organisms in the water column and on the seafloor. The Arctic is also home to diverse land-based mammals, including the caribou, reindeer, fox, musk ox, and polar bear, who all use the frozen sea ice for resting, hunting, birthing, or foraging.

FIGURE 5-15:
Philippe in the Arctic with a boat trapped in sea ice behind him.

Source: Philippe Cousteau

FUN FACT

Contrary to popular belief, polar bears and penguins don't live together. In fact, they couldn't be farther apart. Penguins live in the Southern Hemisphere (mostly in Antarctica but as far north as the Galapagos Islands), and polar bears only live in the Arctic.

Living Under Extreme Conditions: Deep Ocean Ecosystems

Ecosystems can form even in the least likely of places, such as the deep ocean, where the water can be extremely hot or extremely cold, the pressure is intense, and chemicals that are highly toxic to most animals spew into the surrounding water. In this section, we introduce you to four ecosystems you may never have otherwise imagined possible.

Hydrothermal vents

So you like your coffee hot? How about 400 degrees Celsius (750 degrees Fahrenheit) hot?! Yep, that's the temperature of the water coming out of deep-sea hydrothermal vents. These vents or geysers on the seafloor are located on the ridges of tectonic plates, allowing Earth's extremely hot core to heat the seawater and lace it with compounds, such as hydrogen sulfide and methane, which are highly toxic to most life-forms.

It's hot, it's dark, the pressure is the equivalent to ten elephants standing on your head, and the water around hydrothermal vents is full of crazy chemicals, so what could possibly live there? A lot, as it turns out, and it's all thanks to one group of organisms called *archaea,* which are bacteria-like and use chemosynthesis (instead of photosynthesis) for energy. This primary producer forms the basis of a deep-sea food web that includes shrimp, lobster, clams, mussels, tube worms, crabs, fish, and even octopi. Thanks to archaea, life can be diverse and colorful-ish even at the bottom of the ocean (see Figure 5-16).

Deep-sea coral reefs

Did we say coral needs sunlight to survive? Well, we fibbed. More than 3,000 species of deep-sea corals live where the sun doesn't shine, feeding on whatever food scraps float down and past them on their way to the bottom. In addition, like their sun-loving relatives, these corals provide habitats for rich communities of marine life, including marine worms, anemones, sea stars, lobsters, shrimp, crabs, and a variety of fish.

WARNING

If you're thinking, *That's great, I bet these coral reefs are too deep to be disturbed by human activity,* you would be wrong. The biggest threat to these corals is a practice called *bottom trawling,* a form of fishing which crushes the seafloor with enormous rolling wheels destroying these and many other incredible benthic ecosystems in the process. See Chapter 21 for more about this and a horrific before-and-after photo of bottom trawling.

HEY, LOOK, IT'S SNOWING!

Snowing in the ocean? How's that even possible?

Marine snow consists of pieces of organic material that slowly drift down from the surface to the seafloor. It includes pieces of dead and decaying animals, fecal matter, sand, soot, and more. That may not sound appetizing to you, but the organic matter is a sufficient food source to keep some deep-sea ecosystems going.

It can take weeks for this "snow" to reach the bottom of the ocean, and we have yet to encounter a snowman on the ocean floor. Yet.

Cold seeps

Cold seeps are like hydrothermal vents, but instead of pouring out heat, they slowly leak hydrogen sulfide, methane, and other hydrocarbon rich fluids, all of which are used by chemosynthetic bacteria to produce food for other organisms. Due to the pressure in these deep-sea areas, the chemicals come out in the form of a dense brine that lies on the bottom of the sea floor, creating "brine lakes." Along the edges of these brine lakes, you can find mussels living off the archaea and tubeworms that can live up to 250 years. Shrimp and crabs feed on the detritus from the mussels and tubeworms, and then predatory species such as octopus, fish, and crustaceans feed on the shrimp, crabs, and other small creatures.

Whale falls

Not to be confused with a waterfall, such as Niagara Falls, a *whale fall* is an ecosystem that develops around a whale that dies and sinks to the ocean floor at a depth of over 1,000 meters (3,280 feet). All you need for an ecosystem to develop is a source of food or energy. A 40-ton whale carcass certainly qualifies, creating a smorgasbord for creatures that rely mostly on nutrients that drop down from above (see the nearby sidebar about marine snow). When a whale carcass is in shallower water, it can be picked clean in short order by all types of scavengers, but here in deep water, it can take years or even decades to pick those bones clean.

FUN FACT

Whale falls (and the ecosystems they create) are relatively recent discoveries, unknown until the 1970s.

WHALES AS CARBON SINKS?

People rarely think of whales as storing any significant amount of carbon, but they do. Each great whale, for example, sequesters about 33 tons of carbon dioxide from all the plankton it eats. When that whale sinks to the bottom, it carries all that carbon with it, which is then eaten by other organisms which continue to sequester that carbon. Centuries may pass before that carbon will be released back into the atmosphere.

Moving Out and About: Migratory Species

While some animals spend their entire lives in one ecosystem, many ignore ecosystem boundaries, freely roaming across different ones for food, shelter, or to reproduce. For example, you may bump into a shark swimming around a coral reef one day and visiting a sandy shore or a kelp forest the next. Other animals migrate with the seasons or in response to changes in water temperature or nutrients. Here are a few examples of some of our favorite migratory species:

>> Grey whales travel in groups called pods, spending their summers near Alaska, where they fatten themselves up on plankton and then return to the waters off the coast of Mexico in the winter to breed and safely raise their young. This is the longest annual mammal migration on the planet.

>> Leatherbacks (a type of sea turtle) have been known to travel massive distances across the open ocean, such as between Indonesia and Oregon on the west coast of the United States, heading east to forage for food (jellyfish) and then back west to lay their eggs.

>> Orcas are one of the few species of whales that travel freely from hemisphere to hemisphere in search of food. They also travel from Antarctica to the tropics each year presumably to shed their old outer layer of skin which sloughs off easier in the warm water. That is a long way to go for a little exfoliation!

>> Blue marlin travel long distances following the warm ocean currents, feeding on mackerel and tuna that swim near the surface.

>> Arctic terns hold the bragging rights for the longest migration, flying a winding course from their breeding grounds in the Farne Islands in the UK to Antarctica and back — a total of more than 94,951 kilometers (59,000 miles) every year, nearly twice the circumference of Earth.

>> Salmon are famous for their migration from the ocean to the upper reaches of rivers to spawn. After spawning, most die, and the next generation repeats the journey downstream to the ocean and back again.

>> Spiny lobsters use the Earth's magnetic field for their migration. They usually begin to migrate after the first major storm in the fall to deeper waters offshore where the temperature is more stable. What makes this migration so cool is the way they do it. The lobsters "march" in single file in groups ranging from 3 to 200. Thats a long lobster parade.

>> Some eels migrate long distances (who would've guessed?). A team of Canadian scientists used satellite tags to track an adult female eel from the coast of Nova Scotia to the northern limits of the Sargasso Sea in the middle of the North Atlantic — a journey of more than 2,400 kilometers (1,500 miles). Powered only by her tail. #Boss!

Chapter 6

Taking a Deeper Dive: Beneath the Ocean

No discussion of the ocean is complete without mention of what's beneath it — the seafloor — and everything that affects the shape of the seafloor, fuels its deep-sea ecosystems, and forms the land masses that define the ocean's borders. Not to mention the impact the seafloor has on the rest of the ocean, such as creating tsunamis, impacting ocean currents, and more.

In this chapter, we get to the bottom of these fascinating topics as we dive all the way down to the seafloor and even dig below its surface.

Grasping the Basics of Plate Tectonics

Plate tectonics is the theory that Earth's solid crust is divided into sections called plates that slide around on top of a layer of weaker, more pliable rock that forms Earth's upper mantle (see Figure 6-1). In the figure, the plates are part of the *lithosphere,* which includes the crust (land and ocean) and the uppermost part of the mantle (the plates), which is solid rock. The plates ride atop the *asthenosphere* (another layer of the mantle), which is semi-solid rock. These plates bump, grind, and pull away from each other, causing earthquakes, volcanoes, and tsunamis; creating mountains and valleys; forming new seafloor and destroying old; and causing entire continents to move, collide, and separate.

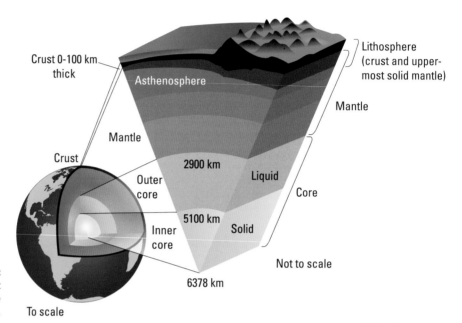

Crust 0-100 km thick

Lithosphere (crust and uppermost solid mantle)

Asthenosphere

Mantle

Mantle

Crust

2900 km

Outer core

Liquid

Core

Inner core

5100 km

Solid

Not to scale

6378 km

To scale

FIGURE 6-1: Earth's crust rides atop the upper mantle.

FUN FACT

Don't expect to feel the earth move under your feet. In a race between a snail and a tectonic plate, the snail wins, no contest. Tectonic plates move about 5 to 7 centimeters (2 to 3 inches) per year.

Earth's crust has seven major plates. Listed from largest to smallest they are: the Pacific plate, North American plate, Eurasian plate, African plate, Antarctic plate, Indo-Australian plate (though some consider this plate to be two different plates — Australian and Indian), and the South American plate. These major plates account for 94 percent of Earth's surface. The other 6 percent is made up of 14 minor plates and 50 to 60 microplates.

How the plates move in relation to one another provides insight into how Earth's crust has been and continues to be shaped and how earthquakes, volcanoes, geysers, and tsunamis occur. Interactions between plates happen along narrow zones referred to as *boundaries,* which come in three types (see Figure 6-2):

>> **Divergent:** Plates pull away from each other allowing magma from the upper mantle to push up through the gap between the plates. At divergent boundaries, you can find rift valleys (as earth from above the boundary drops into the gap between the plates), geysers (as the magma heats any water above it), and ocean ridges and volcanoes (as the magma pushes up through Earth's crust). Divergence is responsible for the ongoing creation of new oceanic crust, as magma pushes up into the gap between the diverging plates, cools, and hardens.

>> **Convergent:** Plates push against each other. When this happens the plates typically buckle upward forming mountains. When two oceanic plates converge, one usually dives below the other one — a process called *subduction*, during which the lower plate is pushed down into Earth's hot mantle and melted (the ultimate recycling). Where one plate dives below another, trenches commonly form in the seafloor. Subduction also occurs where an ocean plate and continental plate converge; the ocean plate typically slides under the continental plate. Convergent plates also give rise to chains of volcanoes.

>> **Transform:** Plates grind sideways against one another instead of pushing together or pulling apart. The San Andreas Fault lies along a transform boundary, where the Pacific and North American plates are sliding past one another, causing frequent earthquakes.

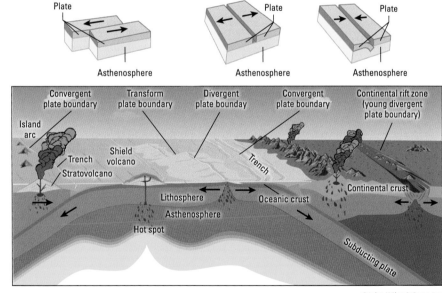

©John Wiley & Sons, Inc.

FIGURE 6-2: Three types of tectonic plate boundaries.

Which leads us to *continental drift*. No, this is not the new *Fast & Furious* film; it is the idea that since the continents and ocean are in constant motion, over long periods of time they drift apart and back together, and apart and back together, and on and on. Whenever they're together, they're referred to as a *supercontinent*. As we mention in Chapter 3, the best-known of these supercontinents was Pangea, which existed between 280 and 230 million years ago. The supercontinent before that was Rodinia, and before that was Nuna (also known as Columbia). Today, many of the continents are moving toward each other again, and will likely smoosh (that's what we like to call it) together to form a new supercontinent to be named Amasia, Pangea Ultima, Aurica, or Novopangea, depending on how the various land masses fit together. Regardless of how it's assembled and what it's called, the general consensus is that most continents will eventually complete their global mash-up within the next 50 million to 200 million years. (Let's plan a party!)

Tracing the Contours of the Seafloor

Seafloor topography (also called *ocean floor topography*) is the study of the land shape and features beneath the ocean. Boring, right? Well, that's what most people mistakenly believed not so long ago, imagining the seafloor to be a flat, barren wasteland. Since then, scientists have discovered that seafloor topography is just as amazing as its terrestrial counterpart. In fact, if you were to take a road trip along the entire seafloor (not recommended, by the way), you'd encounter the following topographical features (see Figure 6-3):

>> **Continental shelf:** A continental shelf is a relatively shallow (up to about 100 to 200 meters or 330 to 650 feet deep) apron around a continent that ends in a comparatively steep slope called the continental slope (covered next).

>> **Continental slope:** The continental slope is a steep drop from about 100 to 200 meters (about 330 to 650 feet) to several thousand meters within a few hundred kilometers. The continental slope extends from the end of the continental shelf (called the *shelf break*) to the beginning of the continental rise. The continental rise has a more gentle slope, which gradually flattens to the abyssal plain (covered next).

>> **Abyssal plain and hills:** The abyssal plain is a flat expanse of seafloor ranging from 4,000 to 6,000 meters (13,000 to nearly 20,000 feet) deep. This plain generally extends from the foot of a continental rise to a mid-ocean ridge and covers about 50 percent of Earth's surface. Abyssal hills are elevations on the deep seafloor that rise from 300 to 1,000 meters (about 1,000 to 3,200 feet) above the seafloor.

>> **Mid-ocean ridge:** The mid-ocean ridge is a continuous range of mountains and underwater volcanoes nearly 65,000 kilometers (more than 40,000 miles) long that makes a winding path around the globe. It's the most extensive mountain range on the planet, and about 90 percent of it is entirely submerged. These ridges form along divergent boundaries, where tectonic plates separate, creating rift valleys and allowing molten rock to spew from the upper mantle, creating volcanoes and mountains.

>> **Volcanic seamounts and islands:** *Seamounts* are underwater mountains mostly formed by now extinct volcanoes. They can rise thousands of feet above the seafloor.

FUN FACT

Although Mount Everest usually gets top billing for being the tallest mountain on Earth at about 8,800 meters (29,000 feet), the Mauna Kea volcano on Hawaii is taller. Its peak is the highest point in Hawaii, rising about 4,000 meters (13,000 feet) above sea level, but its base reaches down another 6,000 meters (10,000 feet) below the surface, making it truly the tallest mountain at about 10,000 meters (33,000 feet) from base to peak.

>> **Trenches:** *Trenches* are long, narrow, deep depressions in the seafloor, accounting for the deepest parts of the ocean. Trenches are formed along convergent boundaries, where one tectonic plate dives beneath another. The bottom of the Mariana (or Marianas) Trench, which is located in the western Pacific Ocean about 200 kilometers (124 miles) east of the Mariana Islands, is the deepest location on Earth at 11,034 meters (36,201 feet) deep. If Mount Everest were resting on the bottom of the Mariana Trench, its peak would still be 2,133 meters (7,000 feet) below sea level.

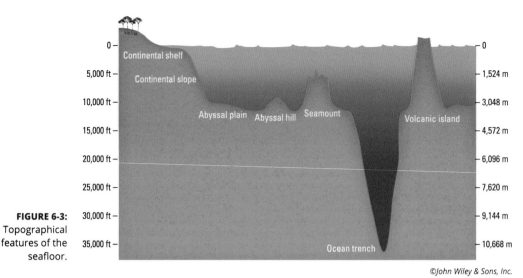

©John Wiley & Sons, Inc.

FIGURE 6-3:
Topographical features of the seafloor.

How does the scientific community know so much about what's at the bottom of the ocean? Keep reading.

Creating the first map of the seafloor

Scientists Bruce Heezen and Marie Tharp created the first accurate map of the seafloor (beneath the North Atlantic Ocean) in 1957. They used *sound navigation and ranging* (sonar) technology — bouncing sound waves off the seafloor, measuring the time they take to travel down and back up, and calculating distances based on those times. (Sonar functions like echolocation used by dolphins and toothed whales to locate food and objects in dark or dimly lit waters.) During the 1950s, computers were just entering the scene, so Heezen recorded the readings, and then Tharp crunched the numbers, by hand and brainpower, to determine the depth at each point. Using this data, Tharp created a map of the calculated depths.

As she was putting the final touches on her map, staring back at her was a mountain range over 16,093 kilometers (10,000 miles) long in the middle of the Atlantic Ocean. When she told Heezen of her finding, he dismissed it as "girl talk" and probably tried to mansplain the impossibility of such a thing to Tharp, but she had just proven the existence of the Mid-Atlantic Ridge. After her male colleague got his head out of his bum and realized Tharp had discovered something incredible (these mid-ocean ridges are the largest geological feature on our planet), they took the finding to the world (sadly, though not surprisingly, as the man, Heezen got all the credit). This finding led to the acceptance of plate tectonics and continental drift.

Tharp and Heezen went on to create the first global map of seafloor topography in 1977. And after a few years and more acceptance of women in science and engineering, Marie Tharp's groundbreaking work has finally been recognized as one of the top discoveries in the entire history of geography. BOOM!

Fine-tuning seafloor maps with better technology

The study of ocean floor depth and topography, technically referred to as *bathymetry,* continues today but with improved technology. Powerful computers and graphics software are now available to crunch the numbers and render detailed maps and even three-dimensional models. Sonar is still in use, but it's been upgraded to *multi-wave* sonar — sound waves fan out from the source in multiple directions to map a larger area of the seafloor at once. The strength of these echoes also indicates the seafloor's composition; for example, soft sediment or hard rock.

Light detection and ranging (lidar) is also used in a similar way to measure the depths and map the contours of the ocean floor by bouncing beams of light, instead of sound waves, off the ocean floor. Scientists are also using satellite data to predict the depth of the ocean based on subtle differences in sea level.

Diving technology has also improved over the years to the point at which ocean explorers can now go all the way down to the deepest parts of the ocean to see it for themselves. As early as 1960, the *Bathyscaphe Trieste* carried Jacques Piccard and Don Walsh (the first humans to descend to the deepest part of the ocean) to the Challenger Deep in the Mariana Trench. The descent took these two hydronauts 4 hours and 47 minutes. They spent only 20 minutes at the bottom and the next 3 hours and 15 minutes to return to the surface. This was the first trip of any vessel, with or without humans, to travel to so deep a point in the ocean. Others have followed in these deep footsteps, including filmmaker James Cameron and Victor Vescovo, who are still arguing over who went deeper (cue the eyeroll).

REMEMBER

Exploring the deep, dark ocean where no one has gone before may seem like just an interesting pastime, but the knowledge gathered from this research has some important practical applications. It provides insight into the relationship between ocean topography and ocean circulation, increases our understanding of ecosystems and environmental processes, and improves management of human activities, including shipping, fishing, and offshore mining — so, yeah, kind of important.

Chipping Away at Ocean Rock and Sediment

All that talk about tectonic plates near the beginning of this chapter may have given you the mistaken notion that the bottom of the ocean is a big flat slab of rock-hard basalt. That's true of the tectonic plates, but a great deal of sediment accumulates on top of those plates.

Rocky seafloors are usually limited to shallow seas, or areas of current or recent volcanic activity where *igneous rocks* (formed by cooling magma or lava) have been in place for a short enough time that sediments have not had time to accumulate. In particular, pillow lava and sheet lava form dense beds of basalt on the seafloor because when a volcano erupts deep in the ocean, the lava flows out slowly due to the pressure and cold water. (That's probably more than you ever wanted to know about rocky bottoms, but there you have it.)

Sediment accumulates fairly quickly on the continental shelf, close to the land from whence it washed into the sea. In contrast, deep-sea sediments accumulate very slowly and are usually fine particles made mostly of *clay minerals* (sediment from the continent that is washed into the deep ocean) or the remains of phytoplankton and the shells of organisms (usually in the form of silicates or calcium carbonate). Because of this, the sediment is thicker either closer to shore or near areas of high biological productivity. The thickness of the sediment layer also depends on how much time has elapsed since the ocean crust first formed at a seafloor spreading site (divergent margin) and whether or not the seafloor is deep enough, cold enough, and with high enough pH (low enough hydrogen ion concentration) to dissolve the calcium carbonate remains that have settled to the bottom.

Sediment comes from other sources, as well, including dust blown into the ocean (you can never escape those dust bunnies) and *precipitates* (small solid particles) that settle on the seafloor. Some of these precipitates come from hydrothermal vents, when heated chemicals released from vents cool. The least amount of sediment is found in areas far from land or near ocean ridges, where new sediment has not had time to accumulate.

Checking Out Deep-Sea Cores

Ever wonder what's in all the sediment and rock at the bottom of the ocean? Probably not, because that would be a little weird. Fortunately, oceanographers and geologists are there to wonder that for us. They take core samples from the bottom of the ocean and study the layers of each core to increase our understanding of Earth's history and prehistory. These layers reveal changes in Earth's atmosphere, climate, and the ocean over millions of years. They can tell us where the continents were at different times, changes in wind patterns, areas of ice sheets, and even changes in levels of oxygen, carbon dioxide, and salinity, not to mention changes in global temperature.

A *core sample* is a cylindrical section of material collected by drilling down into the material with a special, hollow drill bit. If you ever used an apple corer to remove the core from an apple, you know what we're talking about. Using this same concept, scientists drill into solid materials such as land or ice, extract the drill with the sample intact, and then push the core sample out of the drill and study its layers.

A THREAT TO DEEP-SEA ECOSYSTEMS

Many companies are starting to take an interest in deep-sea data, particularly when it provides evidence of valuable natural resources or insight on how to extract these resources. The ocean is home to large oil and natural gas reserves, as well as valuable minerals and gemstones. Many large oil and mining companies have launched underwater projects to dredge, drill, or dig into the seafloor to obtain oil and gas; minerals; valuable metals such as copper, nickel, aluminum, lithium, and cobalt; and even diamonds. Many of these materials are used in everything from beverage cans to batteries, telephones, computers, and cars.

According to the International Union for Conservation of Nature (IUCN), in 2018, the International Seabed Authority (ISA) had issued permits to mine in areas that, combined, would be roughly equivalent to the size of Mongolia. To meet growing demand, many companies want permission to mine in international waters, which is expected to begin in 2025.

Although this is good news for large corporations, it is mixed news for humans in general and bad news for the marine life that call the seafloor home. Deep-sea drilling for oil and gas is pretty much all bad. For example, the Deepwater Horizon oil spill in 2010 released nearly five million barrels (more than 200 million gallons) of oil into the gulf of Mexico. This spill affected more than just the deep-sea ecosystems, devastating beaches, wetlands, and estuaries, not to mention commercial ventures including fishing and tourism. But other forms of mining are more of a mixed bag. Minerals such as cobalt and nickel are used in electric batteries and are usually mined in areas like the Congo in Africa where conflicts break out over who controls access to the minerals or in Southeast Asia where forests are being decimated by mining. Balancing the trade-offs between the human and environmental damage that occurs on land with the potential damage to ecosystems in the ocean is tricky and definitely requires a lot more research before any mining is done. And since the resources of the seabed are considered the heritage of humanity, we also need to work out an equitable way for the value of those resources to be shared with, you know, humanity as opposed to being divvied up among a few big corporations. Or better yet, invest in developing technologies that don't require these minerals in the first place.

3

Sampling the Vast Diversity of Sea Life

Discover what lives in the ocean — from microorganisms too small to see with the naked eye to great big whales, including the largest animal this planet has ever seen.

Take a closer look at microorganisms, some of which are photosynthetic like plants, others that move around and eat like animals, and "tweeners" — photosynthetic organisms that move (weird).

Sample the ocean's flora, including micro- and macro-algae (seaweed), kelp, seagrass, and mangroves.

Check out a variety of *invertebrates* (animals without backbones), such as sponges, jellyfish, anemones, starfish, snails, clams, octopus, shrimp, crabs, and lobsters.

Meet the *vertebrates* (animals with backbones), including a whole lot of fish, a few reptiles, and a wide variety of birds and mammals.

Chapter **7**

Getting to Know the Mighty Microbes

I f strength is in numbers, microorganisms rule Earth, and definitely the ocean. Just look at the stats (all ballpark figures, by the way): By some estimates, a milliliter of water (barely more than a drop) on the ocean's surface contains nearly one million bacteria and ten million viruses. Microorganisms comprise somewhere between 90 and 98 percent of the marine biomass. By last count, the ocean contains an estimated 44 octillion microbes — that's 44,000,000,000,000, 000,000,000,000,000 . . . more that all the stars in the known universe, and more than all the grains of sand on all the beaches on Earth!

But boy are they small — some as tiny as about 1/8,000th the volume of a human cell and about 1/100th the diameter of a human hair. In this chapter, we introduce you to the simplest and smallest life-forms on Earth, a huge majority of which you can't even see without a microscope. We highlight their essential functions, explore their diversity, and explain how they're related to plankton. (We also dig a little deeper into the topic of plankton — the foodstuff of many of the more recognizable ocean creatures — even though plankton includes more than just microbes.)

Meeting the Marine Microbes

Microbes are the most plentiful and diverse life-forms on Earth, but they can be broken down into five basic groups: bacteria, archaea, viruses, protists, and fungi. While some classifications include a separate group for algae, in this section, we stick to the five main groups.

Bacteria

Bacteria are single-celled *prokaryotes* — organisms with cell walls but without a defined nucleus or organelles (specialized structures within the cell). Think of a prokaryote as a bachelor's pad containing all the stuff needed to live but not in any structured arrangement. Bacteria come in all shapes and sizes (see Figure 7-1) ranging from one 1/100th of a millimeter to so big they're visible to the naked eye! Some are primary producers (through photosynthesis or chemosynthesis), while others are primary consumers. Some even eat other bacteria.

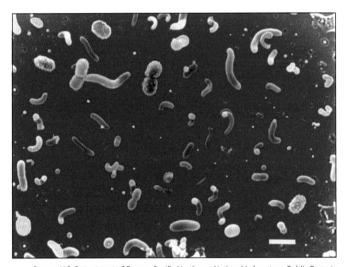

FIGURE 7-1:
A variety of ocean bacteria.

Source: U.S. Department of Energy, Pacific Northwest National Laboratory, Public Domain

FUN FACT

The two largest bacteria are marine species — *Epulopiscium spp.*, found in the guts of Surgeonfish in the Red Sea, and *Thiomargarita namibiensis*, found in marine sediments off the African coast.

The mere mention of bacteria is enough to send some people running for the showers, but bacteria can be beneficial or harmful depending on the bacteria and on your perspective. In fact, the bacteria living in and on your body outnumber the cells that make up your body, and many of them perform essential functions, such as digesting food, manufacturing certain vitamins the body can't make on its own, and even playing a role in fighting infections. Similarly, ocean bacteria serve vital functions such as making food, decomposing waste, and sequestering carbon.

Some bacteria form symbiotic relationships with other microbes and larger organisms. For example, a bacteria known as *Richelia* live inside certain algae cells, while others are specifically cultured inside the organs of some marine animals such as squid and fish.

Marine creatures that glow in the dark (see Figure 7-2) can do so because they have *photophores* — organs that produce light. These organs produce light either through chemical reactions or through the use of symbiotic bacteria that glow in the dark.

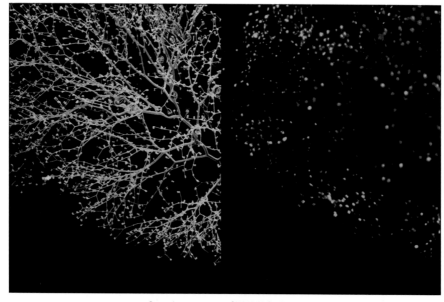

FIGURE 7-2:
The deep-sea coral, *Chrysogorgia*, under regular white light (left) and with bioluminescence (right).

Souce: Image courtesy of NOAA Bioluminescence and Vision on the Deep Seafloor 2015

Archaea

Archaea are an ancient but newly discovered group of microorganisms previously not distinguished from bacteria. Both are *prokaryotes* (single-cell organisms with no nucleus or organelles) as opposed to *eukaryotes* (single- or multicell organisms in which each cell has a membrane-enclosed nucleus and organelles). Bacteria and archaea differ in subtle ways, such as their DNA and the chemical composition of their cell walls.

FUN FACT

Archaea account for about 40 percent of all microbes found in the ocean, and many of them are known as "extremophiles," meaning they can withstand extreme environmental conditions, such as temperatures up to 130 degrees Celsius (266 degrees Fahrenheit) and highly acidic or saline water (up to 30 percent salinity compared to the ocean's average 3.5 percent). They can be found in the hot springs at Yellowstone National Park, in the polar regions, and in the cold, dark deep sea. However, they're abundant in all conditions — so not all of them live life on the edge.

Viruses

Viruses . . . booo! Technically, these microbes aren't even alive (which is why antibiotics are ineffective against viral infections). They're not considered living beings, because they're not cells, and they can't reproduce on their own. They consist of genetic material (either DNA or RNA) along with other molecules enclosed in a *capsid* (a protein shell), and they must hijack a cell (a host cell) to reproduce.

The ocean contains a lot of viruses, like a lot a lot . . . according to at least one estimate, more than 10^{30} — nearly a billion times more than the number of stars in the known universe (but who's counting?).

REMEMBER

Especially after the coronavirus, viruses have become the pariahs of the microbe world, but they, too, provide valuable services. For one, they kill other things, which provides a steady supply of organic matter to scavengers and decomposers. Some beneficial viruses also kill harmful bacteria and may help in killing cancer cells. We could go on, but the point is, as is the case with bacteria, good and bad are relative, so don't judge viruses based on all the bad stuff you hear about them in the news. So . . . yay for viruses . . . we think.

Protists

Protists are eukaryotic microbes that can't be classified as true plants, animals, or fungi. Most are unicellular, some are multicellular, and some cluster together to form their own colonies. They contribute significantly to the ocean's plankton population. While many protists exist as free living forms, many are parasites or symbiotes, relying on other organisms for their survival. Protists are responsible for many diseases, such as malaria in humans (caused by a parasitic protist *Plasmodium*).

Protists are a bit of an oddball group. All of them can seemingly be traced back to a common eukaryotic ancestor, but they're generally unrelated or only loosely related to each other. It's kind of a group where you put anything that doesn't fit in other groups, like uhhh it's not a fungus or an animal or a plant . . . let's just put it in with the other protists. Their classification is always changing as new research comes out. They include the informal grouping *protozoa* (animal-like protists) such as ciliates, foraminifera, and some types of amoeba; slime molds; and algae, such as diatoms, brown, red, and green algae, and dinoflagellates. While many are microscopic, some can grow to over 60 meters (200 feet), such as kelp, a type of brown algae (skip to Chapter 8 for more about algae, diatoms, and dinoflagellates).

Fungi

Fungi are unicellular or multicellular eukaryotes, complete with a cell wall, nucleus, and organelles, but their cell walls are made of a material called *chitin*. Most fungi are multicellular organisms, with different cells responsible for different jobs, such as feeding or reproduction. (Single-celled fungi are called *yeasts*.) Fungi do not photosynthesize; instead, they use a network of tubes known as *hyphae* to absorb and consume organic matter.

SNAILS THAT FARM FUNGI

FUN FACT

Marsh periwinkle snails and fungi that live on Spartina cordgrass have an interesting relationship. The snails feed on the grass, not just to consume it, but to wound it so that fungi can infect the wounds, then grow and multiply. The snails then return to eat the infected grass, including the fungi, which provides them with essential nutrients. (Snails who are fed only the grass, without the fungi, die.)

This practice has been classified as a type of farming, with the snails deliberately creating places for fungi to grow. In return, the fungi gain access to nutritious inner plant tissues and additional nutrients from snail waste (good for them).

When you think of the ocean, "fungus" probably isn't the first word that pops up in your mind. You're more likely to notice them in forests and on that loaf of bread that's been sitting in your cabinet for the past two weeks. However, they hang out in and around the ocean as well — in deep-sea sediments, mangroves, hydrothermal vents, and in the form of lichen in the intertidal zone. *Lichen* is a colony of symbiotic algae and fungi that live in peace and harmony. Some fungi also set up shop inside other living things, such as the guts of animals, or as parasites of certain marine organisms, such as corals, algae, and marine plants.

Similar to bacteria and viruses, fungi can be harmful or beneficial depending on the fungi and the situation. Some are *pathogenic,* meaning they can infect marine organisms and cause disease or even death. But by increasing mortality in microbes, plankton, fish, and even marine mammals, they increase the organic matter available for other living things to consume. Marine fungi also play an important role as decomposers, processing and recycling organic matter, nutrients, and carbon. By converting dead organic matter into fungal biomass, which is then eaten by other creatures, these substances are returned to the food web.

Recognizing the Importance of Microbes

Despite the fact that the vast majority of single-cell organisms and viruses are invisible to the naked eye, we couldn't live without them, nor could anything else on land or in the ocean. In fact, all advanced life-forms we're aware of can trace their evolutionary lineage to these "lowly" beings. Marine microbes are responsible for more than 40 percent of Earth's photosynthesis, sequester about 200 million tons of carbon from the atmosphere every year, produce the food needed to fuel most marine ecosystems, and play a huge role in keeping the ocean clean.

In this section, we highlight a few of the most essential functions performed by microbes that make the ocean habitable for other life-forms.

Feeding the ocean's living organisms

All organisms need energy and matter to survive, grow, and reproduce. For example, a plant uses energy from the sun to convert chemical compounds (energy and matter) into plant cells that form the roots, stem, leaves, and flowers. That energy is stored in the various chemical compounds that make up the plant. An animal that eats the plant can break down those chemical compounds to obtain the energy it needs and the *matter* (chemical compounds again) to form its cells, such as bone, muscle, skin, and so on. The entire cycle involves making and breaking chemical bonds — a process commonly referred to as *metabolism* (see Chapter 3).

For an *ecosystem* (a community of living things) to exist, something needs to bring energy into the system. The beings responsible for bringing energy into an ecosystem are *autotrophs* — organisms that can make their own food from simple inorganic compounds, such as carbon dioxide and water. In contrast, *heterotrophs* need an external food source containing organic compounds (chemical compounds that contain one or more carbon atoms joined with other atoms by a chemical bond) to survive. For example, broccoli is an autotroph, and you are a heterotroph.

Autotrophs can be broken down into two groups — those that use photosynthesis to produce their food, and those that use chemosynthesis. *Photosynthesis* requires energy from the sun to convert carbon dioxide and water into chemical energy and organic molecules needed for growth, whereas *chemosynthesis* obtains the energy it needs from the chemical bonds of inorganic chemicals such as hydrogen sulfide and methane. Either way they do it, the autotrophs are responsible for bringing in the energy (and organic matter) that all other organisms in an ecosystem need to survive. Aww, thanks!

Autotrophs are often referred to as *primary producers* in an ecosystem, because they're responsible for manufacturing the complex organic compounds that provide the nutrients (energy and matter) for all higher life-forms.

Which brings us to microbes — the primary producers that manufacture most of the organic nutrients needed to feed the ocean's other inhabitants (yum, yum). These microbes, along with other primary producers such as seaweed, are eaten by primary consumers (heterotrophs), which are then eaten by secondary consumers (other heterotrophs), and so on, with the organic nutrients (energy and matter) being transferred from one organism to the next.

Energy flow in an ecosystem is often illustrated with an energy pyramid, as shown in Figure 7-3. Energy flows from the bottom up and is gradually reduced as it's used up by organisms at each level.

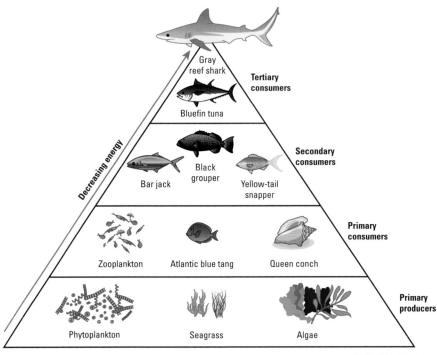

FIGURE 7-3:
A sample
energy
pyramid.

Anchoring food chains and webs

As primary producers, autotrophs are at the bottom of all food chains and food webs. What's the difference between a chain and a web?

» A *food chain* illustrates a sequential transfer of matter and energy (in the form of food) from one organism to the next.

» A *food web* is a group of interwoven food chains that illustrates the complex feeding patterns/relationships among organisms (a.k.a. who eats whom especially when one has a diverse dinner menu). A food web more accurately reflects how ecosystems work.

In the following sections, we explain each in greater detail.

Tracing the food chain hierarchy

A food chain is comprised of producers and consumers, as shown in Figure 7-4. At the bottom, forming the foundation of any food chain, are the *primary producers* — the plants, seaweed, or microbes that make energy and matter available for all other organisms in the food chain. The number one supplier of nutrients and the most important players in most marine food webs are microscopic *phytoplankton* (floating "plants"). Along with a tiny bit of help from seaweed and seagrasses, phytoplankton produce nearly all the organic carbon-based nutrients the entire ocean food web needs (and almost half of all the photosynthesis on our planet).

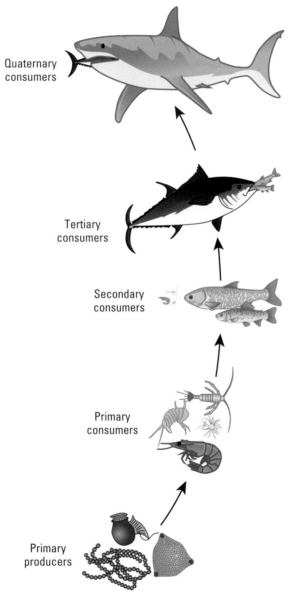

Quaternary consumers

Tertiary consumers

Secondary consumers

Primary consumers

FIGURE 7-4:
A sample marine food chain.

Primary producers

©John Wiley & Sons, Inc.

The *primary consumers* are generally *herbivores* (animals that eat the plants that are the *primary producers*). They include protists, small fish, jellies, zooplankton, invertebrates and even manatees and whale sharks. Minus the manatees and the whale sharks, these animals make a good meal for the *secondary consumers* — *carnivores* (flesh-eating) or *omnivores* (anything-eating). These can be squid, small to medium size fish, and other smallish creatures, which are eaten by, you guessed it, *tertiary consumers* such as dolphins, seals, and penguins.

At the top of the food chain are the *quaternary consumers,* also called *apex predators,* such as sharks and orcas. These are the creatures that eat stuff and generally don't need to worry about being eaten. But there's a catch. Even these perfect predators have to watch out for something — humans. We're not saying that as a "Yay for us," but merely mentioning a sad fact for apex predators.

Introducing intricacy with food webs

A food chain is linear, making it neat and tidy, but in reality, the relationships among producers and consumers are more complex, so biologists use food webs to illustrate these relationships more accurately (see Figure 7-5). Food webs can get messy because animals generally like a little variety in their diets. For example, an arctic cod may eat *zooplankton* (animals that float), *sympagic fauna* (which live all or part of their lives in ice), or *themisto libellula* (shrimplike animals), and the cod may be eaten by a seal, a narwhal, or a variety of seabirds.

The Arctic food web shown in Figure 7-5 contains two *primary producers* — phytoplankton and ice algae, which provide fuel for all the other animals in the system. As the Arctic warms in the spring, *diatoms* (a type of algae with a glasslike cell wall) and other small micro-algae and microscopic protists and tiny animals called metazoans (the sympagic fauna in the figure) trapped in the ice during the winter are released into the ocean as the ice melts. Thanks to the increased sunlight, the diatoms *bloom* (reproduce like crazy), feeding huge swarms of zooplankton which in turn attract fish, seals, walrus, seabirds, giant migratory whales, and other large predators such as orca and narwhals. The spike in algae creates and supports this temporarily rich food web, which then begins to dwindle toward winter as the sea ice re-forms and the migratory animals move to warmer waters.

The food web also includes decomposers, rarely appearing in food web diagrams, that break down waste products and tissues from dead marine organisms into simple organic and inorganic chemical compounds, including water and carbon dioxide, needed by the primary producers. Think of them as nutrient recyclers. They include microbes (such as bacteria and fungi), scavengers (such as crabs and sea stars), and mollusks (such as mussels and oysters). But remember, whether you're an herbivore, omnivore, carnivore, or an apex predator, all your meals are possible thanks to those tiny photosynthesizing (or chemosynthesizing) marine microbes.

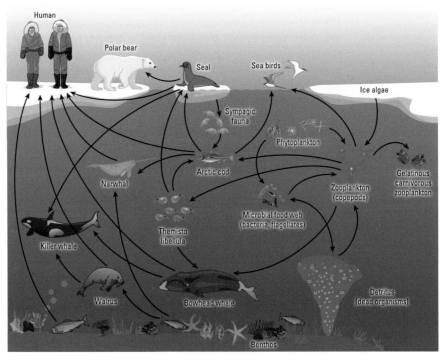

FIGURE 7-5: A sample marine food web.

©*John Wiley & Sons, Inc.*

FUN FACT

Compared to land-based primary producers (mostly plants), microscopic plankton are way smaller, and they lack structural tissues, so they grow and multiply much faster and are generally more nutritious than their land-based counterparts.

FUN FACT

Microbes can also be a source of food while living *inside* another organism. For example, coral polyps have a *symbiotic* (mutually beneficial) relationship with *zooxanthellae* (a type of algae). The algae live inside the coral polyps, which provide the algae a home and protection in exchange for nutrients manufactured by the algae. Similarly, giant tube worms house bacteria within their bodies that provide nutrients in exchange for organic matter. Other deep-sea worms use microbes to help them digest tough organic matter. Hey, what are friends for?

Scaly-foot snails (sea pangolins) have a similar relationship with microbes; the microbes hang out in a specialized gland inside the snail, producing nutrients for the snail. And side note, this snail is super badass! It constructs its shell and the scales on its "foot" from iron sulfide, which makes them slightly magnetic! Unfortunately, they're also the first species to be listed as endangered specifically due to deep seabed mining.

Finally, microbes are also a key component of *marine snow* — the continuous flurry of clumps of organic matter, including waste and decomposing organisms, silt, and microbes that *coagulate* (stick together) and drift down from the surface. As more particles bump into the marine snow particles, the particles get bigger (sort of a snowball effect). Some particles can reach a few centimeters (about an inch) in diameter. (Not quite the size of a delicious snow cone, but the same concept.) Marine snow is an important food source for organisms living in the *aphotic* (deep, dark) zone, including zooplankton, small invertebrates, and fish.

Cleaning up our messes

As nutrient recyclers, microbes are skilled at breaking down organic waste and turning it into something useful. Without microbes, dead organisms and their waste would accumulate, and energy and matter stored in this waste would not be recycled into the food web. In short, microbes help keep the ocean clean and healthy. Fortunately they can also help clean up some of *our* messes, including human waste, oil spills, chemical pollutants, and perhaps even plastic.

In 2010, microbes became a source of study during the BP Deepwater Horizon disaster, when some species were observed breaking down oil released into the ocean. While oil is toxic to most life-forms, some types of microbes live near naturally occurring oil seeps on the seafloor, and consume the hydrocarbons the oil is made of. Different species of marine bacteria, notably *Colwellia*, are highly efficient at removing a small amount of oil from seawater. In some cases, different species may work together to degrade oil over time.

Microbes are also important for breaking down organic waste and pollution in *runoff* (drainage from land to sea). Unfortunately, runoff is so rich in nutrients, it can fuel *harmful algae blooms (HABs)* — overgrowths of algae that harm or kill other life-forms by starving them of oxygen or releasing toxins (see Chapters 8 and 21 for details).

Scientists are also researching microbes and looking into creating genetically engineered microbes that can break down mercury and plastics, which pose serious threats to ocean life (and our own health). Again, this research is still new, but here's hoping.

REMEMBER

We're not saying that microbes can clean up all the messes we make. Despite their efforts in helping to clean up the mess from the BP Deepwater Horizon, none of these species can break down the enormous amounts of oil in a spill in time to prevent damage to the ecosystem. Indeed, oil from that spill continues to negatively impact wildlife. What we really need to do is stop making these messes in the first place while we figure out how to clean up the ones we've already made. See Chapter 21 for more about this.

Looking at the Relationship between Microbes and Plankton

Plankton are drifters — anything that's too small or weak to swim against tides or currents, which includes most microbes that live near the ocean's surface. However, plankton also include larger organisms, such as certain small crustaceans, mollusks, and all jellyfish, along with eggs and the babies of certain animals that will develop into strong swimmers (assuming they're not eaten before they have a chance).

Plankton are generally divided into two groups — phytoplankton and zooplankton — as described in the next two sections.

REMEMBER

Plankton can also be divided into *holoplankton*, which spend their entire lives as plankton and *meroplankton*, which spend only a portion of their lives as plankton. (See the later section "Distinguishing lifers from juvies" for details.)

Phytoplankton

Phytoplankton are tiny photosynthetic organisms that drift. Sometimes described as "floating plants," they lack the complexity of plants and include a certain type of bacteria. Most are different types of single-celled algae such as diatoms, dinoflagellates, and coccolithophores, but some are *cyanobacteria* — photosynthetic bacteria. Diverse, yes, but together they are responsible for a large percentage of global photosynthesis (which creates an enormous amount of oxygen), and they are the primary producers in many ocean food webs. In other words, they matter — *a lot.* (See Chapter 8 for more about marine plants and other photosynthetic marine organisms.)

Zooplankton

Zooplankton are tiny animals that drift (*zoo* = animals and *plankton* = drifter). They're heterotrophs, feeding on phytoplankton, organic matter, and other zooplankton; they can be unicellular or multicellular, microscopic or macroscopic; and they consist of a wide variety of marine creatures. Here's a small sample:

>> Sea jellies.

>> Sea-squirt relatives known as *salps* (barrel-shaped gelatinous organisms that can join together at certain parts of their life cycle to form a colony shaped like a ring or a chain).

» Polychaete worms.

» *Chaetognaths* (pronounced *key*-tog-naths) — predatory marine worms, some with hooked grasping spines near their mouths, some with bioluminescence, some that use neurotoxins. All are weird alien looking things (see Figure 7-6).

» Crustaceans (see Chapter 11) such as amphipods, cladocerans, krill, and copepods (likely the most abundant and diverse group of zooplankton).

» Marine snails such as pteropods whose normal mollusk "foot" has evolved into two winglike flaps for moving.

Photo by Zatelmar. Licensed under CC BY 3.0.

FIGURE 7-6:
A close-up of the head of an arrow worm *(Sagitta).*

Many species of zooplankton are *meroplankton* (plankton for only a part of their lives), such as the larvae of barnacles. (See the next section for details.)

REMEMBER

While zooplankton are weak swimmers, they're relatively mobile compared to phytoplankton. They usually hang out near the surface in the epipelagic zone rarely going deeper than the mesopelagic zone. (See Chapter 4 for more about ocean zones.)

Distinguishing lifers from juvies

Zooplankton can be classified as either holoplankton or meroplankton or, as we like to call them, "lifers" or "juvies":

» *Holoplankton* (the lifers) spend their entire lives as plankton (what a commitment!). These include microscopic organisms such as the following organisms, some single celled and some multicelled:

- *Diatoms:* Algae with silica (glasslike) cell walls.

- *Radiolarians:* Single-celled animals with spiny silica or opal skeletons.

- *Dinoflagellates:* Algae that have two whiplike structures, one for steering and the other for propulsion.

- *Ciliates:* Protists with hairlike structures that enable them to move.

- *Foraminifera:* Known as "armored amoeba" because they have a shell and they move like amoeba.

- *Copepods:* Small crustaceans found in most freshwater and marine ecosystems (see Figure 7-7).

- *Planktonic gastropods:* Groups of free swimming mollusks: pteropods and heteropods.

- *Krill:* Small crustaceans similar to shrimp that form a crucial first step in many ocean food webs.

» *Meroplankton* (the juvies) spend only a portion of their lives as plankton (usually the juvenile stage of certain animals). They either feed on other plankton or survive off nutrients from a yolk sac. Many meroplankton look totally different as adults. Meroplankton include the juveniles of sea urchins, mollusks, sea stars, crabs and lobsters, octopi (see Figure 7-8), marine worms and snails, and some fish.

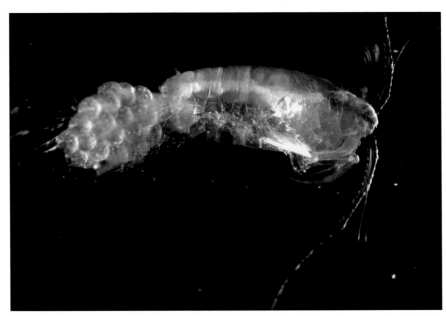

FIGURE 7-7:
A copepod with eggs.

Photo by Matt Wilson/Jay Clark, NOAA NMFS AFSC. Licensed under CC BY 2.0

FIGURE 7-8:
Octopus larva — isn't it cute? You can clearly see that this is a little octopus (juvie) and will get bigger.

Chapter **8**

Sorting Out Algae, Seaweed, and Other Aquatic Vegetation

Marine vegetation is like a gingerbread house — it provides food *and* shelter for other ocean inhabitants, either directly or indirectly. Of course, that's nothing special; vegetation on land does the same thing, and it's far more diverse. But any understanding of marine life begins with an understanding of the plants and plantlike organisms that feed and frequently shelter the inhabitants of many marine ecosystems.

Now marine vegetation may not seem to be the most stimulating topic of conversation, but we assure you that the organisms classified as marine vegetation are a fascinating group. They include algae encased in glasslike shells, algae that can move around like animals, seaweed that can grab hold of the seafloor, bacteria that perform photosynthesis, and even plantlike organisms that live *inside* animals, performing essential tasks in exchange for a place to live!

We know, you can barely contain yourself. We get it, but trust us, settle in, and enjoy the tour; it really is a fascinating world.

All You Need to Know about Algae, and Then Some

Algae is, by far, the most prevalent and diverse marine vegetation. It's sort of like a plant, in that it performs *photosynthesis* (using the sun's energy to convert water and carbon dioxide into carbohydrates), but technically it's not a plant, because it lacks roots, leaves, and a vascular system (a *vascular system* is comprised of vessels, like the veins and arteries in our bodies, that transport nutrients and waste products to and from different parts of an organism). Algae can range in size from microscopic single-celled organisms to multicelled seaweed up to 60 meters (about 200 feet) long, and you can find them pretty much anywhere that's damp and has some source of light.

In this section, we introduce you to *macroscopic* (big) and *microscopic* (teeny tiny) algae and provide insight into a topic that's unfortunately becoming more and more common — harmful algae blooms (HABs).

Go big or go home: Macroalgae (a.k.a. seaweed)

When many people hear the word "algae," they immediately think of pond scum or the slimy green coating that grows inside aquariums. Their thinking is limited to single-celled *micro*algae, but the algae family contains several larger members — *macro*algae, commonly called "seaweed."

Macroalgae are *multicellular,* meaning they're composed of many cells formed into groups that perform different functions for the organism's survival. The smallest macroalgae are only a few millimeters or centimeters in size, while others grow as long as 60 meters (about 200 feet), and they can range in shape from *foliose* (leafy), to crusts, to filaments.

Unlike plants, which have roots, stems, and leaves, seaweed often has a *holdfast* to anchor it to a substrate (for example, a rock or coral), a *stipe* (which resembles a plant stem), and *blades* (which resemble leaves). Some seaweeds have one or more *floats* (gas-filled bladders like balloons) to keep them upright in the water. The entire body of the seaweed (not including the holdfast) is called the *thallus.* Figure 8-1 shows kelp (a type of macroalgae) attached to a rocky bottom. If you look closely, you can see the holdfasts, stipes, blades, and floats.

Most seaweeds are *benthic,* meaning they're attached to the seafloor, but one in particular, sargassum, is free-floating. Seaweeds populate only about 2 percent of the seafloor because most can grow only between the intertidal zone and shallow waters on the continental shelf, where light and temperature conditions are juuuuuuust right (see Chapter 4 for info about the different ocean zones). Some can be selective about the type of substrate they attach to (like dating), while others aren't picky (again, like dating). Different types occupy different habitats, depending on the environmental conditions they prefer such as the salinity, exposure to the sun when the tide is out, nutrients, the strength of nearby wave action, and the presence of grazers (urchins, fish, and other organisms that chow down on seaweed).

FIGURE 8-1:
Kelp attached to the seafloor by its holdfast.

Macroalgae come in three colors — brown, green, and red, making them lovely Christmas presents (just kidding). In the following sections, we celebrate their diversity.

Brown macroalgae

Brown macroalgae include 1,500 species, of which the vast majority are marine (as opposed to freshwater). They generally prefer temperate, not tropical waters, growing in the intertidal zone and shallow coastal waters. They can vary in shape from tiny filaments, to platelike coils, to flat and branching "stems." Many are *perennial* (living for more than a couple years).

LET'S HEAR IT FOR THE MACROALGAE!

Calling macroalgae "weeds" is a grave disservice to these amazing organisms. They provide food and oxygen and add a third dimension to the waters above the otherwise boring, old seafloor, creating a critical environment for a wide variety of organisms to live, hide, eat, and reproduce in. In fact, entire ecosystems form around these so-called weeds, as explained in Chapter 5.

Near the shoreline, kelp forests also help stabilize the sediments of the seafloor and serve as a buffer against waves and storms to prevent coastal erosion. They help filter pollutants from the water and provide a valuable source of food and nutrients for us humans. They're also valuable as a source of biofuel and in the development of medicines, including those used to treat cancer.

Brown macroalgae include kelp and sargassum, both of which can form and support their own ecosystems (see Chapter 5 for details). Larger kelp species can be found offshore, where clear waters and upwellings of nutrients allow some to grow up to 60 meters (about 200 feet) long creating a stunning underwater forest-like world.

Green macroalgae

Most green algae (micro and macro) grow in fresh water or on land; only about 13 percent of them (roughly 1,100 species) grow in marine environments. Most are unicellular, some of which form multicellular filaments or sheets, but as macroalgae they can form long, rubbery strands. Some forms are *coenocytic,* meaning a "leaf" is a single cell with multiple nuclei surrounding a large *vacuole* (like a storage bin for cells, where they put excess waste, water, or nutrients). Some are *calcareous,* meaning they have calcium carbonate in their cell walls, which enables them to contribute to the formation of coral reefs. They tend to be more seasonal than their brown cousins, appearing at certain times of the year, growing rapidly in response to nutrient upwellings, dying out, and then reappearing the following season.

FUN FACT

Certain types of green macroalgae contain toxins to prevent them from being eaten by herbivores. Interestingly, some species of sea slug have adapted to not only tolerate the toxin, but also incorporate it into their own body tissues to protect them against predators. One type of sea slug, which ironically resembles a leaf, known as *Elysia chlorotica,* also keeps the chloroplasts after munching on the algae and uses them to photosynthesize its own food! It's one of the few animals known to be capable of performing this trick. Some experts think that after only a few meals of green algae, these sea slugs can survive a nine-month fast!

Red macroalgae

Red macroalgae are mostly marine and are the most diverse of the three groups with over 7,000 species, their greatest diversity being in tropical regions. While named *red* macroalgae, they come in many colors, from red to brown to yellow, depending on the concentrations of phycoerythrin and phycocyanin and other accessory pigments that give them their color. Most are less than 1 meter (roughly 3 feet) across.

Red algae can form sea lettuce-like structures; structures made of fuzzy, feathery, or long cylindrical branches; or crusts of *coralline algae.* In combination with green and brown algae, red macroalgae can form *algal turfs* — colorful algae lawns or carpets, composed of different types of algae (see Figure 8-2).

FIGURE 8-2: Various types of algae in a tidal pool.

Coralline algae use calcium carbonate in seawater to build their cell walls, making them stronger and more rigid. Some of these algae look like corals or colorful rocks, and they're tough, like a marine cement. Because of this, they're highly important to reef systems; they "glue" different sections of reef rubble together and provide new hard substrate for corals and other organisms to settle on. They can also build new areas of reef solely on their own (like the *algal ridge* — the area of a coral reef closest to the surface, which provides a buffer against waves and storms).

Small, but just about everywhere the sun shines: Microalgae

Microalgae are microscopic, unicellular, photosynthetic organisms that grow individually or in chains everywhere in the ocean where sunlight can penetrate the water. Some float and drift (phytoplankton); some live in or on substrates such as silt, rocks, and the backs of turtles; and some live inside other organisms such as coral polyps. They're broken down into three major groups — diatoms, dinoflagellates, and coccolithophores.

Diatoms

Diatoms are algae with glassy, silica cell walls consisting of two halves that fit together like a shoebox (maybe the earliest form of greenhouses on record!). The shells come in a variety of shapes and may be ornamented with spines, knobs, pores, and grooves (see Figure 8-3), which help to distinguish the different species — by some estimates as many as 200,000 exist, 30,000 of which are planktonic. Collectively, they're responsible for at least 20 percent of Earth's primary production (see Chapter 7 for more about primary producers), so yeah, they're kind of important.

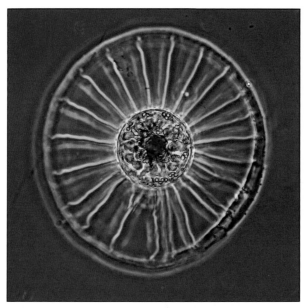

FIGURE 8-3: Wagon wheel diatom — a real beauty!

Courtesy of Dr. John R. Dolan, Laboratoire d'Oceanographique de Villefranche; Observatoire Oceanologique de Villefrance-sur-Mer

These shells are tough, but over time they break down forming a gentle abrasive called *diatomaceous earth,* which is often used in toothpastes and exfoliants. It's also used to treat high cholesterol and support healthy skin, nails, bones, teeth, and hair; to filter beer and champagne (and pool water); and as a natural pesticide to control fleas, slugs, ants, mites, and other insects.

Dinoflagellates

Imagine a houseplant that can walk across the room to get its own drink of water. That's kind of what *dinoflagellates* are — single-celled organisms capable of motion, many of which are photosynthetic like plants. They're characterized by two unique *flagella* (whiplike structures) that propel them through the water, although some lack flagella (maybe they just don't have anywhere to go). Dinoflagellates come in a variety of shapes and can be classified further into armored and unarmored — the armored variety being encased in thin cellulose plates, which may be ornamented with spines or other structures.

Dinoflagellates are usually classified as algae, but some straddle the fine lines that divide the kingdoms of living organisms. They can be *autotrophic* (making their own food), *heterotrophic* (eating others), or *mixotrophic* (able to eat others and make their own food).

Some dinoflagellates are bioluminescent and can light up the ocean at night, sometimes creating beautiful, blue-lit waves crashing against the shore (ahhh, the ambience!). Others, known as zooxanthellae, are important symbiotes of corals, jellyfish, and other heterotrophs, producing food for their hosts in exchange for shelter and protection.

Because they're more mobile than diatoms and less reliant on nutrients such as silica, dinoflagellates have an advantage in environments like the open ocean, which are generally nutrient poor.

Coccolithophores

Coccolithophores (pronounced coco-lith-a-fors) are best known for their calcite scales (*coccoliths*), which help to identify different species. The scales may aid in photosynthesis (sort of like solar panels), defense, stability, or feeding — we aren't really sure, which is kind of cool because it reminds us there is still a lot to discover in this big ol' world.

Coccolithophores were in such high abundance in some areas that, as they died and sank to the seafloor, their scales accumulated and formed chalk and limestone deposits, like the white cliffs of Dover (that's just bananas!!!). When deep-sea cores are taken, the thickness of coccoliths on the seafloor are used to identify periods of warmer seas or higher dissolved carbon dioxide levels (thicker layer = warmer or more CO_2; thinner layer = cooler or less CO_2).

Understanding the threats posed by harmful algae blooms (HABs)

An *algae bloom* is a rapid increase in the population of algae in an aquatic environment. They make the news only when they result in beach closings and massive die-offs of marine life, but they're not all bad. Many are crucial to seasonal food webs. However, some types of algae produce toxins harmful to both marine organisms and humans, or, as toxic or nontoxic algae die and decompose, they deplete oxygen in the water — a condition called *hypoxia* (depleted oxygen) or *anoxia* (no oxygen), as discussed in Chapter 7.

Unfortunately, these blooms are becoming more common worldwide, in both freshwater and marine ecosystems, due to rising temperatures and increased nutrients entering these ecosystems as a result of fertilizers, human waste, and certain chemicals that make their way into the water.

In this section, we explain the different threats posed by harmful algae blooms (HABs) and explain some ways to reduce their occurrence and severity.

Red tide

A *red tide* is a HAB caused by certain types of algae, particularly toxic red dinoflagellates and diatoms, that turn the water rust-red to orange (hence the name). Some red tides are regular natural occurrences, happening once a year or every few years, while others are unpredictable, occurring only when environmental conditions and nutrient levels are just right (or very wrong).

During a red tide, the algae produce and release massive amounts of toxins, such as brevetoxin, saxitoxin, and domoic acid (all nasty stuff). These toxins can cause paralysis; tissue damage; seizures; impaired muscle movement, balance, or orientation; respiratory irritation; and death. Along with the resulting hypoxic or anoxic conditions, these toxins are often responsible for mass die-offs of fish and other marine life (see Figure 8-4) and are extremely dangerous to humans who happen to live nearby as well.

To make matters worse, as plankton ingest the toxins, and then are eaten by small fish, which are eaten by larger fish and other predators and scavengers, the concentration of toxins within an organism increases (called *bioaccumulation*) and at every trophic level in the food web (a process called *biomagnification*). Toxins can also be inhaled by air-breathers, including turtles, marine mammals, and humans. Long after the red tide event is over, concentrations of toxins can continue to biomagnify causing the deaths of larger animals, including seabirds, manatees, dolphins, sea lions, sea turtles, and even whales.

Source: Conor Goulding, © Mote Marine Laboratory

FIGURE 8-4: Beach littered with dead fish and other sea creatures after a HAB.

Brown tide

A brown tide is a HAB caused by the dinoflagellate *Aureococcus anophagefferens* in the eastern United States, but by other species in other regions of the world. This type of algae has a few tricks that give it an edge over other types of algae found in *turbid* (murky) estuarine waters. It tolerates low-light conditions; it can consume organic matter if it doesn't get enough sunlight to make its own food; and it's skilled at breaking down nitrogen and phosphorous (from runoff) when other nutrients are in short supply. Brown tides occur mostly on the EAST COAST of the U.S. and in South Africa, but they're not linked as directly to human activity as are red tides.

Although brown tides may not be as toxic as red tides, they're particularly damaging to marine plants, such as seagrasses, as the high concentration of algae blocks sunlight from reaching the seabed, reducing the ability of these plants to photosynthesize. The algae also clog the feeding mechanisms of filter feeders, such as oysters, clams, and mussels, which can be fatal. The premature deaths of these shellfish can cause *recruitment failure* in some areas (when adult organisms can't spawn to produce new shellfish, so no baby shellfish attach and grow in these areas). This one-two punch to shellfish has been tied to the collapse of some regional shellfish industries, such as in the Peconic Bay area on Long Island.

Pfiesteria

Pfiesteria are heterotrophic dinoflagellates, some of which are toxic. The harmful algal blooms of *Pfiesteria piscicida* in particular (*piscicida* literally means fish killer), are often responsible for the deaths of thousands of fish during a single event. It has a complex life cycle involving many different stages and is thought to produce toxins at some of these stages that can cause ulcers and lesions (potentially fatal), panicked and erratic behavior, air gulping, and drowsiness and lethargy in fish, but this is still being researched.

As tiny, active predators, other Pfiesteria have been documented to swarm and feed on the skin of live fish (one paper described it as essentially stripping the skin off the fish . . . heartless, we know). Pfiesteria species are thought to be most active in waters with excessive nutrients as a result of waste or sewage washing into the system (enriching the estuaries with phosphates and nitrogen), again, thanks to humanity's carelessness.

WARNING

Fish aren't the only victims. Humans exposed to water containing high concentrations of Pfiesteria have been known to suffer a slew of symptoms, including skin lesions, eye and respiratory irritation, headache, gastrointestinal distress, and neurocognitive difficulties, including impaired memory and learning. Geez.

HAB prevention

Harmful algal blooms probably can't be prevented entirely, but we humans can take steps to reduce their frequency and severity. Most HABs are caused by a combination of warm water, sunlight, and fertilizers, sewage, and certain chemicals in runoff, so we need to address the factors we can control. Here are a few preventive measures we can take to minimize HABs:

» Use fertilizers smarter and more conservatively (on both farms and lawns).

» Improve wastewater treatment.

» Properly maintain septic systems and replace them when necessary.

» Address problems with sewage being dumped down storm drains.

» Reduce use of detergents detergents, soaps, and shampoos that contain phosphates.

>> Restore wetlands to slow the flow of chemicals into the ocean.

>> Stop global warming.

Shoring Up the Shoreline with Mangroves

A *mangrove* is a woody tree or shrub that lives along tropical or subtropical coastlines within reach of the tide. (The term also applies to habitats formed by these trees and shrubs.) Approximately 80 different mangrove species have evolved to live in silty, oxygen–poor, *saline* (salty) environments.

While most plants can't tolerate high concentrations of salt, mangroves are specially adapted to being flooded daily with salt water. Their roots are specialized to filter out most of the salt. Any excess salt that happens to make it past the first line of defense is directed to the leaves, where it's secreted through certain pores or glands and washed away by the rain, or the leaves are shed, along with the salt.

But salt isn't the only challenge mangroves have overcome; they also need the ability to anchor themselves to the shoreline where sediment tends to be very loose, as well as obtain enough oxygen for respiration. (Yes, plants need oxygen, too, for all their living tissues, including their roots, which is why house plants die when they're overwatered.) How do they do it all? Again, it's roots to the rescue! Mangroves anchor themselves to the shoreline through their vast network of roots. To obtain the oxygen they need, some mangroves have aerial roots called *pneumatophores* that grow upright out of the sediment and water, sort of like tree snorkels. Others have *prop roots* (see Figure 8-5), which grow from the trunk or the branches of the tree straight into the sediment, so that parts of the root remain above the waterline.

Source: Claudio Contreras-Koob – www.wildcoast.org

FIGURE 8-5:
Prop roots in action.

How are mangroves like mammals? They bear live young. Unlike other plants, which produce dormant seeds that don't sprout until they reach soil, mangrove seeds sprout while they're on the tree. They may grow a foot tall before they fall off and join the family or float away to start their own families. How cool is that?!

Like the kelp forests and seagrass meadows, mangroves create important habitats for coastal species both above and below water, all while protecting shorelines, sequestering loads of carbon, improving water clarity and quality, and contributing to local economies (see Chapter 5 for details).

Not Your Typical Lawn: Seagrass

Unlike seaweed, *seagrass* is a true plant, which just happens to live underwater. They have vascular tissues, leaves, roots, and *rhizomes* (just like lawn grass) — modified stems that grow horizontally under the substrate and produce new plants (roots and shoots). They also reproduce like flowering land plants, but instead of being pollinated by insects, birds, and wind, the pollen is carried by tides and currents.

Some seagrasses have symbiotic relationships with nitrogen-fixing bacteria. The seagrass introduces small amounts of oxygen into the sediment (which may be low in oxygen), enabling the bacteria to produce nitrogen compounds that serve as fertilizer for the seagrass.

About 60 species of seagrass include eel, manatee, surf, shoal, paddle, and turtle grasses, which are distinguished by leaves (blades) of different shapes and sizes. Most leaves are flat and ribbonlike (see Figure 8-6), but some can be cylindrical or oval. Seagrasses are particularly diverse in the Indo-West Pacific Ocean, especially around the tropical waters of Australia.

Seagrasses, like kelp and mangrove trees, can be ecosystem engineers, forming seagrass meadows that support a diverse community of marine life. Seagrasses also play a valuable role in sequestering carbon, cleaning water, stabilizing the seafloor, and protecting coastlines from waves, tides, and storms (see Chapter 5 for details).

Sadly, at least 1.5 percent of seagrass meadows are destroyed every year, globally. This amounts to two football fields lost *every single hour*. The worst culprits are runoff (which can make the water too murky for the seagrass to obtain the sunlight it needs) and dredging or improperly anchoring boats (which tears entire plants up from their roots). Removal of patches of seagrass causes fragmentation of the meadows, making them more susceptible to being entirely ripped from the seafloor by natural events, such as storms. Invasive species, global warming, and reductions in populations of predators that prey on the herbivores that eat the seagrass also pose threats. And if all that isn't enough to wipe them out, these stressors make seagrass more susceptible to bacterial and fungal infection.

FIGURE 8-6:
Seagrass.

Checking Out What's Growing in the Salt Marshes

As we explain in Chapter 5, salt marshes (Figure 8-7) are muddy coastal areas that are regularly doused in ocean water during high tides. The diversity of plant life that grows in salt marshes is limited by the fact that the plants that grow here must be salt tolerant, but how tolerant they need to be varies based on their relative location in the salt marsh.

Early in the formation of salt marshes, plants like glassworts and cordgrass are the first to take hold, because they're tolerant of salt, submersion, and soil that's *anoxic* (containing little or no oxygen). Glassworts are succulents with fleshy stems and scalelike leaves that grow in salt marshes, on beaches, and among mangroves. Cordgrass is a tough, wiry grass that spreads vegetatively with *rhizomes* (underground stems) that sprout new plants. These rhizomes do wonders to stabilize the soil.

FIGURE 8-7:
A salt marsh.

These types of plants serve as "pilot" plants (totally trend makers) creating soil conditions that enable other plants to grow, especially higher up in the salt marsh where the ground is generally flooded only during higher-than-average tides. Plants commonly found in the high marsh include salt hay grass, spike grass, black grass, salt marsh asters, and sea lavender.

Salt marshes are vitally important ecosystems that provide food to countless species, as well as critical nursery habitat for juvenile animals, including many species that are vital to coastal economies. They are also important nesting grounds for birds, and they filter runoff and help prevent coastal erosion. Unfortunately , we are losing seagrass marshes at an alarming rate (notice a trend?).

Chapter **9**

Getting the Lowdown on Simple Invertebrates

nvertebrates are animals that have no backbone. Yes, they're spineless creatures, but that doesn't make them any less interesting. In fact, they include such a fascinating array of marine creatures we had to break them down into three big groups — *simple invertebrates* (lacking heads and limbs for the most part), *mollusks* (characterized by soft, squishy bodies, though most have shells), and *crustaceans* (animals that wear their skeletons on the outside). In this chapter, we cover the simple invertebrates, and in the next two chapters, we shed the spotlight on mollusks and crustaceans.

Each of these three groups includes a diverse population of sea creatures. The "simple" invertebrates alone include a variety of sponges, jellies, anemones, coral polyps, starfish, sea urchins, sand dollars, sea cucumbers, animals that look more like plants but can walk or "fly," and a respectable collection of worms, yes, worms. Be prepared to witness a parade of fascinating beings, many of which might strike you as aliens from another planet.

Sponges and Other Holy Creatures: The Porifera

Sponges are sedentary aquatic animals with soft, porous bodies, which explains why they're technically referred to as *poriferans* — pore bearers. The pores lead to a network of canals and tiny chambers through which the sponge pumps water containing food and oxygen (see Figure 9-1). Cells inside the sponge extract the oxygen and the nutrients (from the food) and eliminate wastes that are then pumped out of the sponge through a large opening called the *osculum*. This method of extracting nutrients from water is why sponges are classified as *filter feeders*. Some sponges also have *cyanobacteria* (photosynthetic bacteria) living in their cells, which produce food for the sponge in exchange for protection and a place to live.

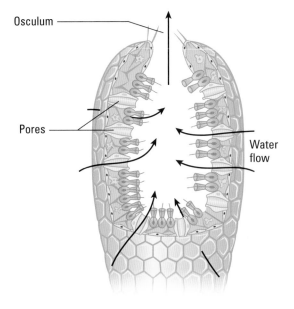

Osculum

Pores

Water flow

FIGURE 9-1: Water and nutrients flow through a sponge.

©John Wiley & Sons, Inc.

Most sponges are *hermaphrodites* (both male and female), and they can reproduce sexually or asexually (how's that for options?). To reproduce sexually, one sponge releases sperm that floats to another sponge of the same species to fertilize its eggs. The hatchlings (larvae) float around and swim (a little) until they latch onto an object on the seafloor and start to grow. Sponges can also reproduce asexually through *budding* — a piece of the sponge detaches from the main body and starts to grow somewhere else.

All adult sponges are *sessile,* meaning they're permanently attached to the sea-floor, which must get boring, especially given the fact that some are estimated to live longer than 2,300 years!

Sponges are thought to be the original reef organisms, because they evolved before corals. Some sponges even use calcium carbonate, as corals do, to create a structure for support. Another similarity with corals is that sponges are a crucial building block of ecosystems, serving as food for some fish and sea turtles and as shelter for small species of fish and shrimp that live inside them.

The number of sponge species is estimated to be 5,000 to 10,000, most of which are marine but some of which live in fresh water. To make them more manage-able, the scientific community has broken them down into four groups, as pre-sented in the following sections.

WARNING

Most household sponges contain plastics. They break down as they're used and long after they're discarded, and some of the microplastic fibers eventually make their way into the environment. Look for eco-friendly alternatives, such as reus-able cotton cloths and sponges made of natural fibers such as cellulose (plant fibers). Real sponges from the ocean are a good option as well, as long as they're harvested responsibly, because they can grow back.

Calcarea

Members of the class *Calcarea* are known as "calcareous sponges" because they have small, sharp, brittle, *spicules* made of calcium carbonate embedded in their bodies. These spicules provide structural support and self-defense. Calcarea include roughly 675 species found exclusively in the ocean, predominantly in the shallows, but a few in the deep ocean. They generally resemble vases or tubes, or groups of tubes joined together (see Figure 9-2).

Demospongiae

Demospongiae is the most common and diverse class of sponges, accounting for about 81 percent of all known sponge species, including the large barrel sponge shown in Figure 9-3. Demosponges contain *spongin,* a fiber that gives them their flexibility and makes them very attractive for commercial use. They're common in most marine environments from the tropics to the poles, and in some freshwater environments, and they come in a variety of shapes, sizes, and colors.

FUN FACT

With a diameter of up to 2 meters (about 6 feet), the giant barrel sponge is the largest of its kind. Imagine having that sitting in your shower.

FIGURE 9-2:
Yellow calcareous sponges with a diamondback tritonia nudibranch in the foreground.

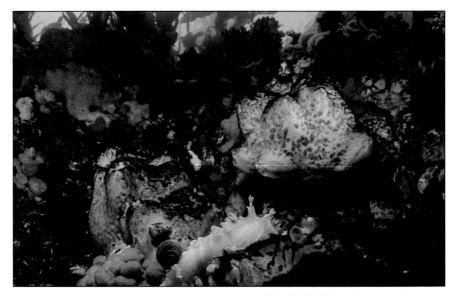

FIGURE 9-3:
A large barrel sponge with a rock hind grouper sitting inside.

Hexactinellida

Members of the class *Hexactinellida* are more commonly known as *glass sponges* because they're translucent and delicate, like those glass knickknacks your grandparents were always warning you to be careful not to break (see Figure 9-4). These sponges are exclusively marine and are common in the deep ocean and in colder waters, such as around Antarctica.

NOAA Image courtesy of the NOAA Office of Ocean Exploration and Research,
Deep-Sea Symphony Exploring the Musicians Seamounts

FIGURE 9-4:
A delicate glass sponge.

Hexactinellida (try to say that three times fast) includes roughly 623 species, but this number is expected to rise as more of the world's ocean is explored. Just recently, a sponge thought to be a hexactinellida measuring over 3.5 meters (11.5 feet) was discovered by deep-sea rovers off the coast of Hawaii. (Guess you're not so big after all, giant barrel sponge!)

Homoscleromorpha

Homoscleromorpha is a relatively limited class of sponges, with fewer than 100 species. They have tiny spicules made of silica and are usually small and coral-like or *encrusting*, which means they fasten onto rocks or other objects and often grow next to and on top of one another to cover the object. Like calcarea, they generally hang out in shallow water, but they've also been spotted in the abyssal zone and in underwater caves.

Jellyfish, Anemones, and Other Notable Cnidarians

Cnidarians (pronounced ny-*dare*-ee-uhns) are animals with unique stinging cells, known as *cnidocytes,* but that's not all they have in common. To belong to the club, they must have radial symmetry, like a pie (see Figure 9-5). If you cut it into wedges, each wedge looks pretty much the same. We humans have bilateral symmetry, so our left and right sides look pretty much the same, but if you cut us into wedges, they won't look anything alike.

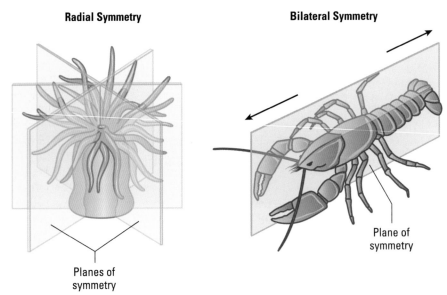

Radial Symmetry

Bilateral Symmetry

Planes of symmetry

Plane of symmetry

©*John Wiley & Sons, Inc.*

FIGURE 9-5: Radial versus bilateral symmetry.

The cnidarian life cycle consists of two stages — the *polyp* stage, during which it's attached to the seafloor (like an anemone or a coral polyp) and the *medusa* stage, when it's free-swimming like a sea jelly. Some cnidarians, including corals and anemones, exist only as polyps, whereas others exist mostly as medusas.

In all forms, cnidarians have an internal cavity with an opening at one end (the "mouth"), which is usually surrounded by stinging tentacles. During the polyp stage, the mouth and tentacles face upward, and during the medusa stage, they face downward. Much of the body is composed of a jellylike substance known as *mesoglea,* which explains the term "jellyfish."

Positioned around the mouth and tentacles are stinging cells called *cnidocytes,* which are the signature characteristic of cnidarians. Inside each cnidocyte is a capsule-shaped organelle called a *nematocyst,* which contains a thin, coiled tube that may or may not have a harpoonlike structure at the end. On the surface of the nematocyst is a triggerlike mechanism called a *cnidosil.* When an enemy or prey activates the cnidosil, the coiled tube is discharged and either pierces the skin of the target (or wraps around it) and then expels its venom, which kills or immobilizes the target.

Cnidarians are generally divided into four classes (groups): scyphozoans, hydrozoans, anthozoans, and cubozoans. Some classification schemes recognize one or two other groups, such as staurozoan, but we're going to stick with the four traditional classes, as presented in the following sections.

Scyphozoans

Scyphozoans, considered the "true" jellies, represent about 200 species including moon jellies, jellyfish, and sea nettles. They have both a polyp and medusa stage, but the medusa stage is more dominant. While many drift along with ocean currents and are considered zooplankton, others are more active and can swim in all directions by pulsing their bodies. Most jellies use long, stinging tentacles to stun or kill prey, before bringing the food back into their mouths. While the sting is deadly to most creatures, some small fish have adapted resistance to the venom and use the tentacles of jellyfish for protection against predators.

Some jellies also have photoreceptors that can sense whether an environment is light or dark. Some types of golden jellyfish living in marine lakes in Palau have a symbiotic relationship with algae that live in their tissues; the jellyfish provide protection and access to sunlight, while the algae provide nutrients for the jellyfish. Every day, hundreds of thousands of these jellyfish follow the movements of the sun, so that their algae have enough sunlight to produce food for them. Other types of jellyfish, such as upside-down jellyfish, have a similar relationship with algae, and "sunbathe" in the shallows to feed.

FUN FACT

The largest jellyfish in the world, the lion's mane jellyfish, has a *bell* (main body) more than 2 meters (roughly 6.5 feet) wide, and tentacles trailing over 36 meters (roughly 120 feet). True story: While we were scuba diving off Vancouver, Philippe pulled me (Ashlan) out of the path of one of these bad boys. I hadn't seen it floating my way, and it missed me by only a few inches. That would not have been a pleasant experience.

Hydrozoans

Hydrozoans, commonly called "hydras," are similar to scyphozoans in that they have both a polyp and medusa stage, but with hydras, the polyp stage is more dominant. Many hydras form colonies composed of many individuals performing different functions; for example, some are specialized for catching prey and others for reproduction. Colonies are typically small, reaching only about 15 centimeters (6 inches) in height, and the polyps are usually tiny.

The most notable of these colonies is the Portuguese man-o'-war (see Figure 9-6), which consists of several different types of polyps. The topmost polyp forms that gas-filled bladder that enables the colony to float above the surface. Other polyps are responsible for capturing prey, feeding, digestion, and reproduction. How's that for teamwork?

FIGURE 9-6:
A Portuguese
man o' war.

Source: Keith Ellenbogen – www.keithellenbogen.com

Some hydras are solitary and, in their medusa stage, are often confused with scyphozoans, but the medusa of hydras have a muscle shelf projecting inward from the periphery of the bell, which makes them unique.

Other hydrozoans, such as fire corals (not actually coral), produce a skeleton made of calcium carbonate and are known as *hydrocorals.* And man, their stings hurt.

Another famous creature in this group is the "immortal jellyfish" — a hydra that can somehow revert back to its juvenile phase when injured, stressed, or old, making it biologically immortal, though many die before they can turn back the clock. While really cool, they're starting to concern ecologists because they appear to be successful invaders — an invasive species that never dies . . . what a nightmare!

Anthozoans

Anthozoans, literally "flower animals," include corals and sea anemones. Many anthozoans form large colonies, and exist only as polyps, similar to hydrozoan polyps, but larger and more complex. Like other cnidarians, many anthozoans have tentacles to sting/stun and capture prey. Sea anemones are capable of contracting their tentacles back into their bodies when not feeding, and some can move short distances to avoid predators.

Sea anemones are famously used by clownfish for protection, hiding among the anemone's tentacles without being stung (see Figure 9-7). In exchange for protection, the clownfish delivers food and cleaning services. So, the million dollar question — how does the clownfish manage not to get stung? Well, it has a coating of mucus about four times thicker than that of other fish, some of which it may collect from the anemone. According to one hypothesis, this mucus is inert or contains a substance that prevents the nematocysts from firing.

FIGURE 9-7:
A clownfish makes itself at home in a sea anemone.

Source: Cristina Mittermeier – www.sealegacy.org

Other types of anemone deliberately attach to crabs — their backs, shells (in the case of hermit crabs), or claws. For the crabs, we like to think of it as accessorizing . . . very stylish. Some, such as the blanket hermit crab, go all in, using the anemone *as* its shell. However they choose to co-exist, the crab provides transportation and food scraps, while the anemone provide camouflage and defense.

Some polyps are strictly sessile. For example, coral polyps stake out a single spot and build the intricate structures that make up coral reefs. Each coral animal is a community of tiny polyps. The shape, structure, and consistency of corals vary from branches, to plates, to boulders, some hard, some soft, depending on the type of coral. (See Chapter 5 for more about corals and coral reefs.)

Cubozoans

Cubozoans, such as the box jellyfish, are characterized by their cube-shaped bells. This group includes some of the most venomous animals in the world, such as the sea wasp (a type of box jellyfish and the largest and most deadly of the group, allegedly the most venomous sea creature), and the Irukandji jellyfish (about the size of your fingernail). The stings of a cubozoan can cause extreme pain, and their toxins can be fatal to humans, attacking not only the skin, but also the heart and nervous system. (Check out Chapter 22 for a list of the ten deadliest sea creatures.)

Some have complex eyes, despite not having a brain (like ours), and are relatively agile and strong swimmers (for jellyfish, that is). Most have four main tentacles used to capture and kill prey. They're also unique for having what is believed to be courtship behavior! I wanna kiss you all over.

Ctenophora (Comb Jellies)

FUN FACT

When is a jelly not a cnidarian? When it's a *ctenophore* (pronounced ten-uh-four), commonly called a comb jelly. While they resemble jellyfish (jellylike and translucent) and most of them have tentacles, they lack the stinging cells of cnidarians. Instead, they have sticky *colloblasts* used to capture prey. Comb jellies have eight "comb rows" of fused *cilia* (like tiny hairs) that run along their sides and beat or flap in rhythm to propel them through the water. Some are bioluminescent, and when their cilia beat they scatter the light, like a prism, generating a rainbow of colors.

The phylum *Ctenophora* contains only about 200 known species, but they inhabit most ocean zones from pole to pole and from the surface to the deep ocean. Most of them live in the water column, closer to the surface, but a few species hang out nearer to the seafloor.

Starfish, Urchins, and Other Famous Echinoderms

Echinoderm means "spiny skin" — a fitting name for the members of this phylum, which include sea stars, brittle stars, sea urchins, sea lilies, and sea cucumbers. They're predominantly benthic animals, and can be found at most depths, from shallow seas to the deep seafloor (see Chapter 4 for more about ocean zones). Echinoderms generally exhibit radial symmetry, but some, such as sea cucumbers, are bilaterally symmetric. Echinoderms have an internal skeleton, made of calcium carbonate plates called *ossicles,* just beneath their skin that help them maintain their shape. The spines characteristic of urchins and many other echinoderms protrude from these ossicles. Some echinoderms are covered with *pedicellariae* (which resemble tiny pincers, like Edward Scissorhands) to keep their skin free from parasites and other freeloaders.

Thanks to an internal vascular system, echinoderms are able to move around (albeit pretty slowly). Water enters into the system via a *madreporite* (a perforated plate), usually positioned at the top of the animal, and is pumped through a network of channels. Tube feet can be found at the end of some of these channels. As water is pumped into these tube feet, they extend outwards. Each tube foot has longitudinal muscles that can swing the foot in different directions. Using this combination of hydraulic and muscle power, they can move their feet back and forth to slowly walk or climb.

FUN FACT

Starfish feet don't suck; they stick. Although the bottoms of the feet look like tiny suction cups, they're actually equipped with a pair of glands that secrete two substances — one that acts as an adhesive and another that acts to release the adhesive.

Asteroidea (sea stars)

And now for the star of our show — *Asteroidea,* more commonly known as sea stars or starfish, though they're not really fish. These fascinating creatures are emblematic of ocean life, and they're *everywhere,* from rock pools in the intertidal zone and all the way to the deep ocean. With nearly 2,000 species by some estimates, they're also incredibly diverse.

Most sea stars have five arms, but some have as many as 40. Some are bumpy, some have spines, and some are smooth. On the underside of these arms are the tube feet, which enable the sea star to move in any direction (see Figure 9-8). And if they lose an arm or two, no problem — they can just grow a new one. In some cases, the arm itself can grow into a whole new sea star!

Also on the underside, smack dab in the middle, is the mouth, which comes in handy because most sea stars are voracious predators, typically feeding on invertebrates, including clams and scallops. They can actually pry open a stubborn clam to dine on the tender meat inside, but not with its mouth. The sea star has an unusual way of eating: It pushes its stomach outside its body onto the clam (or other unfortunate creature), releases digestive juices that dissolve the animal's flesh, and then absorbs the soupy mixture. (That's like something you'd see in a horror movie.)

FIGURE 9-8:
The underside of a sea star showing its tubed feet.

Source: NOAA/NOS/NMS/FGBNMS; National Marine Sanctuaries Media Library Licensed under CC BY 4.0

Ophiuroidea (brittle stars)

The class *Ophiuroidea* includes brittle stars, basket stars, and serpent stars (Ophiuroidea translates to "serpent stars"). In particular, brittle stars look like sea stars on a diet; their arms are relatively skinny, curly, and flexible, and are attached to a roundish central disk (see Figure 9-9). Some types have spiny arms. They're known as brittle stars because when they were first collected for science, their bodies were fragile and broke very easily while being studied.

Brittle stars feed mostly on organic debris, or they filter feed, although some will hunt for small animals. To trap plankton and organic material, they wave their arms around, and sticky mucus between their spines captures their food, which is then passed to their mouth. Others use their tube feet to move food on the seafloor to their mouth.

FUN FACT

Brittle stars are able to sense their prey by using chemical and light sensors on the ends of their arms. Some are even able to change colors to blend in with their surroundings to avoid being eaten, while others can emit a flashing light to distract predators — they'll drop a bioluminescent arm for the predators to chase while the brittle star "runs" the other way.

Echinoidea (sea urchins and sand dollars)

Members of the class *Echinoidea* include sea urchins (Figure 9-10a) and sand dollars (Figure 9-10b), which couldn't be more different. One looks like a porcupine, and the other like a flying saucer. Heart urchins are sort of a cross between the two, some looking more like sand dollars and others more like sea urchins. However, they all have more in common than you may think. All echinoids have a rigid external skeleton called a *test*, made of tightly interlocking plates. They also all have spines, although they may be long on some and barely noticeable on others. Finally, most have a structure known as *Aristotle's lantern* (five teeth arranged in a circle).

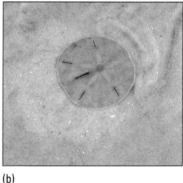

FIGURE 9-10:
Sea urchin and
sand dollar. (a)   (b)

In addition to their differences in appearance, sea urchins typically hang out on hard, rocky surfaces or on kelp, which they love to eat, whereas sand dollars and heart urchins generally prefer the sand. Also, sea urchins use their long, pointy, and sometimes venomous spines for self-defense, whereas sand dollars and heart urchins use their short spines more to bury themselves in the sand.

Most echinoids are herbivores that graze on marine vegetation, while others are omnivores, feeding on organic debris either in the sediment or suspended in the water.

Crinoidea (sea lilies and feather stars)

Crinoids are marine animals that look more like plants. They're even named like plants — feather stars (Figure 9-11a) and sea lilies (Figure 9-11b). Sea lilies look more like flowers or palm trees, complete with a holdfast or *cirri* (like roots), a stalk, and feathery arms that look like petals or leaves. Feather stars lack the stalk making them look more like shrubs. As a group, crinoids are quite colorful.

Sea lilies generally live in deeper waters, where they anchor themselves to the seafloor and grow tall — up to 1 meter (3 feet). Without a stalk, sea feathers are shorter and tend to live in shallow water, though they're also found in the deep sea. Sea feathers are more mobile than sea lilies, using their arms to "fly" through the water or crawl short distances along the bottom.

Crinoids are considered "suspension feeders," meaning they filter food from the water using the tube feet on their feathery limbs, and like many other invertebrates they usually feed on organic debris or plankton. Most crinoids are found at depths greater than 100 meters (330 feet). Those that live in the shallows prefer to take refuge during the day and emerge at dusk to feed. Like other echinoderms, they can regenerate lost limbs, a great party trick.

FIGURE 9-11:
A feather star perched on a sea fan and a sea lily photographed in the deep ocean.

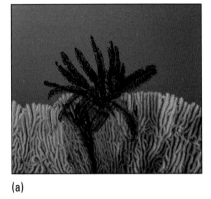
(a)

(b)

Holothuroidea (sea cucumbers)

Holothuroidea is a class of long, leathery invertebrates, numbering about 1,250 species, that come in a wide variety of sizes, shapes, and colors. Some look like our garden cucumbers or squash, others look like big caterpillars, and one even resembles a piglet, aptly called a sea pig. They range in length from a few centimeters (about an inch) to nearly one meter (about 3 feet), and their skin can range from smooth to rough to bumpy to spiny (see Figure 9-12).

Sea cucumbers are generally omnivorous, feeding on waste, organic debris, algae, and other small organisms. They collect their food using sticky modified tube feet (like tentacles) positioned near their mouths. Some use their tube feet to pick up food, while others consume the sediment or sand, extract the nutrients, and poop out the sand and sediment. They have a similar method for breathing, but at the other end — they draw water into their "lungs" through their *cloaca* (their posterior orifice, to say it nicely), extract the oxygen, and expel the water. In other words, they breathe through their butt.

FUN FACT

In addition to having an appearance that would scare the bejesus out of most predators, sea cucumbers have a couple other unique self-defense tactics. Some eject sticky threads when in danger, sort of like Spider-Man, and make their escape as the predator tries to free itself. Others expel a portion of their internal organs out of their butts, leaving a snack that keeps the predators busy while they escape . . . or maybe the predators are so grossed out that they lose their appetite (no, not really).

Source: Schmidt Ocean Institute – www.schmidtocean.org

Squirmy Wormies: The Annelids

Annelids are segmented worms, including earthworms and leeches. They're segmented in the sense that their bodies consist of multiple repeating divisions. Annelids don't have a hard skeleton; instead, they have a "hydrostatic" skeleton consisting of a fluid-filled compartment that runs the length of the body, providing it with structural support. Many annelids also have small bristles/hairs on the outside of their bodies that help them dig and move around. Most of them play an important role in breaking down and recycling organic matter to provide nutrients for communities in which they live.

Fair warning, some annelids are venomous.

REMEMBER

The annelid club has a membership of about 9,000 species, which traditionally have been divided into three classes — polychaeta, oligochaeta, and hirudinea, as presented in the following sections. However, the classification system for worms is a work in progress, so these class divisions are likely to change. Our primary goal in this section is to increase your awareness of the diversity of these creatures.

Polychaetes

Eight thousand of the 9,000 or so species of annelids are *polychaetes* (Latin for "having much hair"). Each segment of these worms has a pair of paddle-shaped appendages covered with tiny bristles used for eating, breathing, and moving. As a class, they comprise 40 to 80 percent of marine *infauna* (animals that live in the sediment), and in some locations thousands of them crowd into an area as small as one square meter (talk about a crowded square!).

Most polychaetes live in the sediment of the upper continental shelf and the deep ocean floor, and they include sandworms, feather-duster worms (which look a lot like feather stars), tube worms, and sea mice (which, at first glance, could easily be confused for heart urchins).

They're often divided into two informal groups based on how much they move. The sessile types, called *sedentary polychaetes*, reside in burrows or inside tubes they construct themselves (from sand, rocks, calcium carbonate, proteins, or whatever's accessible); they remain in place, only poking their heads out from time to time. Some are filter feeders, such as fan worms and Christmas tree worms (see Figure 9-13), trapping food in brightly colored, sometimes featherlike tentacles. These worms also have light-sensing organs, enabling them to detect potential predators and retreat back into their tubes or burrows when they feel threatened.

Source: Cristina Mittermeier – www.sealegacy.org

FIGURE 9-13: Christmas tree worms.

The mobile polychaetes, known as *errant polychaetes*, may also live in burrows or tubes, but they have the option to move around. They often reside in the open ocean or in shallows where they crawl between rocks and crevices or on and between the shells of other organisms. Many are active hunters with menacing jaws and teeth they use to capture and kill their prey. Others are ambush predators, hiding in burrows until their prey comes close and then snatching it and dragging it back into their lair. Some are more amicable, feeding on dead animals or algae.

Oligochaetes

Oligochaetes (Latin for little or few hairs) include earthworms (mostly) and a few marine species. Marine worms in this group are usually tiny and transparent and move around like earthworms, using the tiny bristles on their skin to shift grains of sand. Most are found in the intertidal zones underneath rocks or macroalgae, or in the deeper ocean zones.

Like polychaetes, oligochaetes have light-sensing organs to tell light from dark. And, like most other worms, they have sensory organs to help them "smell" their way to food — mostly dead organic matter. Both terrestrial and aquatic oligochaetes can breathe through their skin, but some aquatic species have specialized gills.

Hirudinea

Hirudineans, which are less segmented than other annelids, include leeches and tiny aquatic parasites. Most have at least one "sucker," which enables the worm to attach itself to the host or to another object, whereas others have a sucker at each end to help them move (like an inchworm).

Many marine leeches feed on blood; they attach to their prey (shorebirds, water birds, fish, sharks, rays, and so forth), make a small incision, and release an anticoagulant (to prevent the blood from clotting) and a histamine (to cause the blood vessels to dilate, thus increasing blood flow). However, while leeches are usually considered to be bloodsucking parasites, some are carnivores, gulping down their prey whole. Regardless of how or what they eat, leeches rely on their prey's chemical scent to locate it.

Chapter **10**

Getting Mushy over Mollusks

ollusk is derived from the Latin word "mollis," meaning "soft," which accurately describes the bodies of these creatures — soft and squishy. Chances are good that you've already encountered a few mollusks on land — snails and slugs. You may have met a few more at the beach or along the shores of a nearby lake or stream or at your favorite seafood restaurant — maybe an oyster on the half shell, a clam in your chowder, a scallop, or even a squid (often referred to as "calamari," to get people to order it). Or maybe you bumped into a few during a visit to your local aquarium.

Unfortunately, most people meet these fascinating creatures only outside their natural habitats and never get to know just how diverse, colorful, and interesting they are. Well, that's about to change. This chapter provides a backstage pass to the wonderful world of mollusks, a couple of which just happen to be our favorite sea creatures of all time.

Meet the Mother of All Mollusks

Paleontologists have good evidence to suggest that mollusks evolved during the Cambrian period, between 541 and 485 million years ago (see Chapter 3 for more about the prehistoric ocean). What they haven't been able to figure out is what the first mollusk was or what it looked like. So, they've done what all good scientists do — they made an educated guess (technically speaking, a hypothesis). In this case, they imagined what the first mollusk probably looked like, and they came up with a model for it called the *hypothetical ancestral mollusc*. They even have an acronym for it — *HAM*.

While we may poke fun at this scientific process, it's actually very useful for our discussion, because HAM gives us a great way to explain what makes a mollusk a mollusk. After all, looking at a snail and an octopus side-by-side, you'd never guess that they're members of the same family. HAM helps explain their similarities (see Figure 10-1):

>> A soft, squishy body, unsegmented with bilateral symmetry (see Chapter 9 for more about radial and bilateral symmetry)

>> A *radula* (toothed tongue)

>> A *mantle* (a fold in the body wall that lines and produces the shell)

>> A muscular foot (expressed as arms or tentacles in some mollusks)

>> A *coelom* (a fluid-filled cavity)

Although most mollusks have a shell, it's considered an optional accessory. Some have an internal shell, or they have a shell during a certain stage in their lives. Some have a reduced shell you may not even notice. Others have no shell at all.

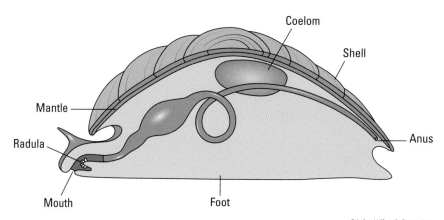

FIGURE 10-1: The hypothetical ancestral mollusc (HAM).

©John Wiley & Sons, Inc.

Gastropods: Putting Their One Foot Forward

The largest group in the mollusk family (accounting for about 80 percent of all mollusk species), is the class *Gastropoda,* which literally means a stomach with a foot on it. The name strikes us as more than a little demeaning, not to mention inaccurate — after all, snails and most slugs have easily recognizable heads and a complex anatomy, including a liver, lung, heart, kidney, a primitive brain, and, yes, a stomach. In addition, they're incredibly diverse. Gastropods are one of the few groups that have members living in all three main habitats — land, sea, and fresh water. As a group, they also have a varied diet. Depending on the species, they may be herbivores (plant eaters), scavengers, carnivores, or even internal parasites.

These diverse creatures come in many sizes, some with shells and some without, but they all have a single "foot" they use mostly for locomotion. To move, they excrete mucus and contract their muscles in waves to slide over the surface of hard objects. Think of it as skating over ice that you create in front of you as you move in a certain direction. In this section, we break the family of gastropods down into two groups — snails (with external shells) and slugs (mostly without external shells).

Snails

Snails are gastropods that carry their homes on their backs. Many types can retract completely into their homes headfirst and seal them shut with an *operculum* (a thin, rigid disk attached to the foot of most snails). This door, which is a little softer than the shell, keeps predators out and moisture in, giving them a much wider travel range and enabling them to survive dry conditions.

REMEMBER

When you hear the word "snail," you probably think of the archetypical snail with the spiral shell, but snails are far more diverse, as you're about to discover. In the following sections, we introduce you to several different types of snails, some of which you may never have imagined being snails.

Abalones

Abalones, which account for about 60 to 100 gastropod species, are edible gastropods covered by a bowed shell that looks sort of like a flattened helmet with a line of holes along one side used for breathing and excreting wastes (see Figure 10-2). The inside of the shell is lined with mother of pearl, which is commonly used in jewelry and to adorn other objects. Abalones also produce pearls, although they rarely receive much credit for doing so.

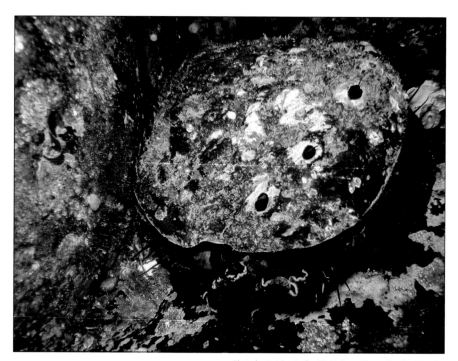

Photo by Lt. John Crofts, NOAA Corps. Licensed under CC BY 2.0

FIGURE 10-2:
An abalone.

Conches

Conch is a general name that refers to medium-to-large mollusks that have a very angular shell with a wide lip. These are the shells you hold up to your ear to hear the sound of the ocean or you blow into to produce a loud sea call (which takes a lot of practice, trust us). Conches have some of the most creative names, such as queen conch (shown in Figure 10-3), dog conch, and horse conch, the latter of which is a giant predatory gastropod. When it eats something, well . . . it just looks darn nasty — like a massive tongue with no mouth or body slurping something up.

FUN FACT

Some conches can produce pearls, which are regarded as some of the rarest and most valuable pearls in the world (even though they're not a "true pearl" but a calcareous concretion from the conch). Who knew?

Cone snails

Cone snails account for nearly 500 species of venomous, predatory sea snails each equipped with a harpoon containing a large collection of toxins. When its prey gets close to a cone snail, the cone snail fires its harpoon, piercing and then paralyzing (or killing) it, which it then swallows whole. (See Chapter 22 for ten deadly sea creatures.) Cone snails are cone shaped (of course) with a long, linear *aperture* (opening), as shown in Figure 10-4.

FIGURE 10-3:
A queen conch.

Photo by G. P. Schmahl, NOAA FGBNMS Manager. Licensed under CC BY 2.0

FIGURE 10-4:
A cone snail.

Source: Prof Jamie Seymour – James Cook University

Cowries

Cowries have a smooth, glossy, domed shell with a long narrow opening. Their mantle has two special folds that the cowrie can wrap around the outside of the shell. These folds are covered with *papillae* (fleshy, hairlike projections) that may serve as camouflage or enhance respiration. Cowry shells are highly prized among shell collectors and have even been used in days long past as a form of currency.

Limpets

Limpets have a thick, conical shell (see Figure 10-5) and a strong, muscular foot. The shell is difficult for predators to grab hold of or crack, and the limpet uses its foot to draw itself tightly against rocky substrates, making it really hard to pry from the surface.

FIGURE 10-5: A limpet grazing on algae; notice the scoured rock behind it where it has eaten all the algae.

Source: NOAA's Fisheries Collection, Mandy Lindeberg, NOAA/NMFS/AKFSC. Licensed under CC BY 2.0

FUN FACT

Slipper limpets often live in stacks, like upside-down saucers. Where they are in the stack influences their sex. A large female is usually at the bottom with a small young male on top. When the male releases his sperm, it drops down to fertilize the female's eggs. Limpets between the top and bottom may be male or female.

Periwinkles

Periwinkle is a name used loosely for all small freshwater and marine snails, but true periwinkles are marine snails that can be found in the ocean, on rocky shores, in estuaries and mud flats, and on the roots of mangrove trees. These sweet little gastropods have a small spiral shell built to withstand the constant beating of waves. They are also the favorite food of many seabirds — ducks, in particular, *love* them!

Whelks

Whelks are predatory snails that use their radula to drill through the armor of their prey (such as the shell of a mussel) to eat the soft flesh inside. They can also use their foot to take hold of a bivalve (such as a mussel or clam) and use their own shell as a wedge to pry open the shell. The whelk's shell is similar to that of a conch, so look back at Figure 10-3 to get a pretty good idea of what a whelk looks like.

Sea slugs and sea hares

Sea slugs and sea hares are basically homeless snails. At some point in their evolution, they lost their shells, though many species have a tiny internal shell or a small external shell that barely covers their body. What's cool, though, is that without shells to protect their soft bodies, they've evolved several interesting and effective self-defense mechanisms — camouflage; the ability to collect venom from what they eat and use it on their enemies; bold, bright colors to make them look venomous (regardless of whether they truly are); and the ability to detach a part of their bodies (as some reptiles do) and grow them back later.

Now, if you're thinking, "Slugs, yuck!" you may be surprised to find out that more than 3,000 species of sea slugs called *nudibranchs* (which means "naked gills") include some of the most diverse, colorful, and ornate creatures in the sea (see Figure 10-6a–c). If there were a beauty contest for marine animals, a nudibranch would probably claim the crown year after year.

Some nudibranchs look like fluorescent hedgehogs, some look like glow-in-the-dark slugs on acid, and some look like gorgeous neon Flamenco dancers. All in all, they're beautiful underwater gems. Seriously, we could do a whole giant coffee table book on dazzling nudibranchs.

Sea hares are very similar to sea slugs, but they have large tentacles that resemble the ears of hares (rabbits), and they have winglike structures they use for swimming. Some also squirt inky substances that look like a smoke screen but are thought to stimulate food receptors in predators, so they try to eat the substance while the sea hare escapes.

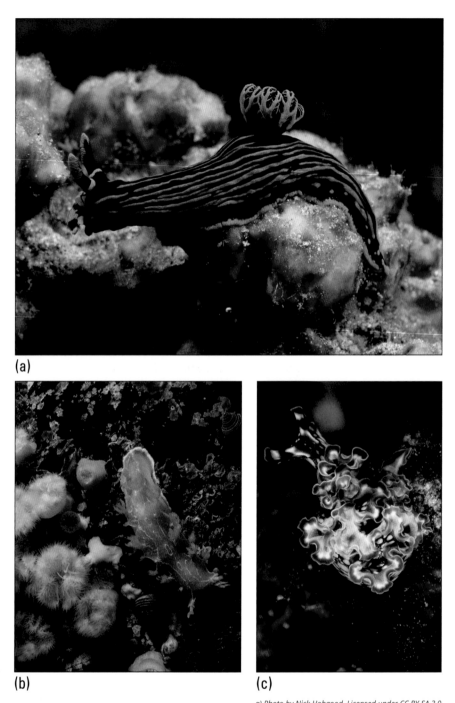

(a)

(b)

(c)

a) Photo by Nick Hobgood. Licensed under CC BY-SA 3.0
b) Source: Cristina Mittermeier – www.sealegacy.org
c) Source: Romona Robbins – www.RomonaRobbins.com

FIGURE 10-6:
A small gallery of nudibranchs.

Bivalves: Parts One and Two

Bivalves are headless mollusks with a hinged, two-part shell, sort of like castanets. They're very diverse, boasting a membership of more than 15,000 species divided into four main groups — clams, oysters, mussels, and scallops — generally based on where they live and the shape of their shells. Clams prefer an *infaunal* lifestyle, buried in the sand or silt, whereas mussels, oysters, and scallops live a more *epifaunal* lifestyle (on or above the seafloor).

Regardless of whether they're in-ground or above-ground, they have specialized gills that serve two functions — to extract food and oxygen from the water. They play a key role in cleaning the water and, fortunately, live in or near just about every body of water — oceans, lakes, ponds, rivers, and streams.

FUN FACT

Bivalves lead interesting sex lives. Clams and other burrowers tend to be *dioecious,* meaning they're male or female their entire lives. Many scallop species are *serial hermaphrodites,* able to produce male and female gametes (sperm and eggs) their entire lives. Various oyster species are *protandric hermaphrodites* — males in their youth and then changing to females and staying that way as they get older (you go, girl). European oysters are *rhythmical consecutive hermaphrodites,* changing back and forth over the course of their lives.

Clams

If you've been to the ocean, you probably had a close encounter with a clam, although you may not have noticed, because they usually bury themselves in the sand for safety. To breathe and feed, they extend a long double-tubed *siphon* just above the surface of the sand that works sort of like a snorkel. One tube draws in water containing food and oxygen, and the other expels water, carbon dioxide, and wastes.

Unlike oysters and scallops that move around on the seafloor at least a little, most clams restrict their movement to digging down and digging out. They use their mollusk foot less as a foot for locomotion and more as a shovel for digging. They also have two powerful adductor muscles located on opposite ends of the shell to close their home up tight when they feel threatened and to seal in the moisture in case they get stranded on shore.

Most clams are filter feeders, although some supplement their diets by establishing symbiotic relationships with algae (in shallow water) or chemosynthetic bacteria (in deep water or areas that are high in sulfides). Giant clams are somewhat famous for their symbiotic relationship with algae. In fact, they've been referred to as "greenhouses for algae" — directing beneficial light to the mantle, where the algae hang out, but because too much light can be harmful to algae, the clam's filter also redirects harmful light wavelengths away from the algae, making said algae feel right at home. Maybe this is why the giant clam is the largest of its kind. It can grow to be more than 1 meter (3.2 feet) across and weigh more than 181 kilograms (400 pounds)! (See Figure 10-7.) And honestly, they are a beautiful sight to behold.

FIGURE 10-7:
A giant clam.

Source: Kevin Lino NOAA/NMFS/PIFSC/ESD. Licensed under CC BY 2.0

Oysters

Oysters are awesome. You can eat them, build reefs with them, harvest their pearls to make beautiful jewelry, and even use them to clean the ocean. In fact, they're top-of-the-line water filters! A single oyster can clean about 8 to 11 liters (2 to 3 gallons) of water an hour, and they never get tired or break down. It's no wonder conservationists and habitat restoration projects use oysters all over the world to quickly and naturally clean up dirty water and restore ecosystems. (Check out the "Billion Oyster Project" off NYC in the Hudson River by visiting www. billionoysterproject.org).

Oysters are actually a subclass of saltwater clams but with irregularly shaped shells, which is how you can tell an oyster from a clam, a mussel, or a scallop. Those other bivalves have smoother shells with both halves matching pretty closely. Oyster shells are irregularly shaped and rougher. Their outer surface looks more like a bad concrete job but they can produce the most beautiful pearls on the inside. And its what's on the inside that really counts!

FUN FACT

You can tell the color of a pearl before you even open the oyster by looking at its "lip" — its outer edge, because they form pearls out of the same material (*nacre*) they use to create their hard inner shells. These inner shells are the source of "mother of pearl," which is used to create jewelry and adorn other objects.

Mussels

Mussels hang out in freshwater lakes, streams, creeks, and rivers, along with the intertidal zone, where ocean meets land. They've been cultivated in Europe since the 1400s for food (steamed in a white wine sauce with a side of French fries, please). In marine environments, mussels are often sessile, anchored to rocks in high–flow areas with their byssal threads, as explained in the nearby sidebar. They're filter feeders, playing an important role in keeping the water clean.

Scallops

Scallops have two fanlike shells joined by a straight hinge, which, along with the muscle inside that joins the two shells, enables them to "swim" short distances. Okay, it's not really swimming; it's more like a clapping motion that propels them through the water, but it gets them where they're going. If you've never seen a scallop swim, head to YouTube right now and watch a clip. Scallops are also unique in that they have two rows of light-sensing organs, like eyes, arranged along either side of the scallop's opening (see Figure 10-8).

FIGURE 10-8:
A scallop has
"eyes."

Cephalopods: Head and Tentacles Above the Rest

While *gastropod* means "stomach on a foot", *cephalopod* means "head on a foot", but with these mollusks, the *pod* (foot) has evolved into many prehensile arms/tentacles, which may be equipped with suction cups, hooks, or gooey mucus to catch prey and perform other functions. Yes, we're talking octopus and squid, and their close cousins, the nautilus and cuttlefish. Most cephalopods have 8 to 10 arms, but some (such as the nautilus) have as many as 90. Now that's a lot of handwashing!

If you look at a bivalve and a cephalopod side by side, you'd never imagine they were in the same family. Not even close! Every cephalopod has a sophisticated brain, three hearts, good eyesight, a system of jet propulsion, prehensile arms, a sharp beak, and (in most species) an ink sac for self-defense — whereas a clam is a hunk of flesh sealed in a shell that can attach itself to rocks and sips through a straw (not that there's anything wrong with that).

Most cephalopods lack the distinct shell that's characteristic of most mollusks. One exception is the chambered nautilus, which has a well-developed shell with air-filled chambers to keep it afloat. The cuttlefish, which kinda looks like a nautilus without an external shell has an internal, elongated, saucer-shaped shell called a *cuttlebone,* which is often sold at pet stores as a calcium source for birds (which seems kinda wrong). The squid has a long, thin, internal shell called a *pen.*

In this section, we introduce you to the four most common members of the cephalopod family.

Octopi

If aliens exist on this planet, they're octopi or octopuses (both spellings are acceptable, by the way). The octopus (see Figure 10-9) is regarded as one of the most intelligent creatures in the sea, and *the* most intelligent invertebrate on Earth thanks to its large brain. In fact, the brain-to-body ratio of the octopus is the highest of all invertebrates and greater than that of many vertebrates. It even has a group of nerves that act brain-ish for each arm, enabling the octopus to move them independently. The octopus is also a tool user and can learn and remember. (Full disclosure, they're Philippe's favorite animal)

FIGURE 10-9:
An octopus.

Source: Cristina Mittermeier – www.sealegacy.org

Moving on to the body, an octopus has eight arms, each of which has two rows of suckers used to capture and hold prey and to stick to smooth surfaces. The arms lead to a skirt, in the middle of which is their mouth (beak). With three hearts, they have lots of love to give. One heart pumps blood through the body, while the two small hearts pump blood to the gills. Their bodies are very malleable, allowing them to squeeze into super tight spaces — as long as their beak fits, the rest of them fits too, bringing a whole new meaning to the phrase, "If I fits, I sits."

Compared to bivalves, their sex lives are ultra-conservative. Males remain males and females remain females their entire lives. As soon as the male passes sperm to the female to fertilize her eggs, the female becomes a devoted, die-hard mother . . . literally. For example, the giant pacific momma octopus lays her eggs and attentively watches over them, keeping them clean, aerated, and protected for up to ten months, during which time she doesn't leave and doesn't eat. She usually dies shortly after her eggs hatch. Octopus fathers don't fare much better — they often die after mating (talk about deadbeat dads!).

FUN FACT

An octopus will always beat you at a game of hide and seek. They're able to change their color and texture to match their surroundings to a T. But the mimic octopus has everyone beat; it can even change its shape to impersonate other creatures such as a flounder, a lionfish, a sea snake, or even a tube worm. Hey, do you guys hire out for parties?

Honestly, we could write an entire chapter or even a whole book about octopi, given how fascinating they are. Just look at how adorable the dumbo octopus is (see Chapter 4), and the mating ritual of the argonaut octopus is something we just can't talk about in a book for family audiences. Ahhhhh, so much to say, but so little time.

Squid

Squids (see Figure 10-10) look a lot like octopi, but they're different in many ways, including the following:

>> An octopus is smarter than a squid, but squids are better swimmers.

>> An octopus has a roundish body, rectangular pupils, and eight arms, whereas a squid has a triangular body with a fin on either side, round pupils, eight arms, and two longer tentacles (with suction cups only at the tips).

>> The arms of an octopus are more flexible than those of a squid, enabling them to walk around and to hold and move objects.

>> A squid has a rigid internal structure, called a pen, that runs along its mantle and provides support; an octopus does not.

>> Octopi generally hang out on the seafloor eating crustaceans and other benthic prey, while squid prefer the open ocean, feeding on shrimp and small fish.

>> A squid's self-defense mechanism involves expelling a cloud of ink that serves as a smoke screen, whereas an octopus relies more on camouflage or squeezing its body into a hollow object or crevice (though, in desperate situations, an octopus can ink too).

>> Octopi reproduce as partners and attend to their eggs for up to a year until they hatch, whereas squids mate in large groups and leave their fertilized eggs attached to rocks or corals to fend for themselves.

>> Octopi are generally solitary, whereas squids may live alone or in groups.

FIGURE 10-10: Squid hatching from clusters of eggs.

Source: Cristina Mittermeier – www.sealegacy.org

Squids range in size from about 16 millimeters (less than 1 inch, and so cute) up to 22 meters (about 72 feet, and terrifying) when stretched out.

Some cool species of squid include the glass squid — almost fully transparent except for its eyeballs (though its eyelids act as an invisibility cloak); the vampire squid, which can turn itself inside out to avoid predators; and the Humboldt squid, which can pulse its body with flashing red and white bioluminescence. Yowsa!

THAT'S ONE BIG CALAMARI!

The giant squid is about 8 meters (26 feet) long, but with its tentacles stretched out, it may reach 22 meters (72 feet) in length. These massive creatures live in the deep ocean, and scientists still don't know much about them. Most of what's known has been gathered from studying carcasses that have washed up on beaches or been brought in by fishing boats.

Based on the limited information available, we know that they eat shrimp, fish, and other squids. We also know that they engage in defensive epic battles with whales and sharks (that like to eat squid), based on the fact that whales and sharks have been observed with what look like giant squid hickeys all over them.

Because they live in the deep sea, giant squid have giant eyes. We're talking BIG, as in largest in the animal kingdom — about 25 to 30 centimeters (10 to 12 inches) in diameter (the better to see you with, my dear!). Researchers also think that giant squid live only about five years, meaning they must grow like weeds, and that they mate only once, so they'd better make it count.

However, while the giant squid may be the longest, it may not be the largest. The colossal squid is shorter but weighs twice as much. One colossal squid on display at the Ta Papa Museum of New Zealand tips the scales at 490 kilograms (just over 1,080 pounds), while an average Giant Squid weighs in at around 275 kilograms (606 pounds). The beak of the colossal squid is the largest of all among mollusks, and their eyeballs are about the size of soccer balls.

Even with their massive size, the giant and colossal squid are the preferred prey of the deep diving sperm whale. And (fun job), some scientists study the undigested beaks of these squid in sperm whale stomachs to gather additional information about the species. That would be one colossal and very smelly day at the office.

Cuttlefish

Cuttlefish, also known as cuttles (no, not cuddles, although they look kind of cuddly), are sort of a cross between a squid and an octopus but with a more compact body. Like an octopus, a cuttlefish has a big brain and is a master of camouflage. Like a squid, it has eight arms and two longer tentacles and its head and body are tapered, more like a torpedo. Cuttlefish are unique in that they have an undulating fringe running along their sides and a cuttlebone to help with buoyancy, which enables them to hover (see Figure 10-11). Another unique feature is their pupils, shaped like a "W," which enables them to see in front of and behind them at the same time.

They tend to live in deep water during the winter and in the shallows over the summer months, and they live only one or two years, dying shortly after mating.

The giant cuttlefish lives in the waters around Australia. These large cuttles can grow to have a mantle length of about 50 centimeters (20 inches) and weigh around 9 kilograms (about 20 pounds). That's a lot to cuddle! They come back every season to the same rocky shores of southern Australia and mate, lay eggs, and then die. When the next generation hatches, they head off into the world (not much is known about where they go or what they do), but they always return to the same area to mate, lay eggs, and perish (cue up "Circle of Life" from *The Lion King*).

Nautilus

If a snail, a shrimp, and an octopus had a baby together, it would look like a nautilus (see Figure 10-12). The nautilus has a spiraled shell like a snail, but it's sectioned off into chambers containing air to make the nautilus buoyant, enabling it to hover in the water. As the nautilus grows and expands its shell, it creates new chambers. It has a face like a shrimp and arms like an octopus — actually about 90 tentacles that are grooved and secrete mucus to capture food and hold onto stationary objects when resting. Compared to octopi and squid, the nautilus doesn't have the greatest vision, relying more on its sense of smell to find food.

FIGURE 10-12:
A nautilus.

Source: Nautilus Macromphalus, depth 110m, Laurent Ballesta – laurentballesta.com

They're *nocturnal* (active at night), making daily migrations up and down the water column. They live much longer than the other cephalopods (up to 20 years). Unfortunately, their shells are highly prized and because they don't reach sexual maturity until they're 10 to 15 years old (and even when they do, females lay only about ten eggs max), their population has declined significantly and will require a long time to bounce back. And because they need a hard shell to survive, they're also threatened by ocean acidification. Thankfully, they're a protected species, though poaching still happens.

FUN FACT

The nautilus is considered a living fossil, because it has changed very little over the 500 million years it has been around (in some form). Today, they live in the waters of the Indo-Pacific, hovering above reefs at depths of about 100 to 300 meters (330 to 990 feet). They can't go much deeper, because the pressure would crush their air-filled shells. *Not* a good way to go.

Chapter **11**

Wearing Their Skeletons on the Outside: Crustaceans

What do shrimp, crabs, lobsters, krill, barnacles, and sand fleas all have in common? They're all crustaceans, meaning they're crusty, like a loaf of bread — hard on the outside, soft on the inside. Technically speaking, they have a hard *exoskeleton* (external support structure) composed mostly of chitin. Of course, they have more in common than that, which we get to in a minute, but that gives you a general idea of what a crustacean is.

Although we touch on what makes a crustacean a crustacean and explore similarities between different species, we also celebrate the diversity of these fascinating and kind of freaky-looking sea creatures by highlighting a handful of them. The crustacean club (phylum) has a membership of more than 68,000 species that range in size from 0.1 millimeter long (barely visible to the naked eye) to the size of a Japanese spider crab — nearly 4 meters (12 feet) long. Most have appendages, but some don't. Most move around, while some are sedentary creatures. Some graze, while others hunt. Some live only a couple years (at most), while others have been known to live for 50 years in captivity.

In this chapter, you don't get to meet them all (unfortunately), but we introduce you to a cross-section of the crustacean family, so you can appreciate how diverse they really are and meet some of the more remarkable members of the family.

What Makes a Crustacean a Crustacean?

To qualify as a crustacean, an animal must meet the following minimum requirements:

>> A tough exoskeleton composed mainly of chitin, which must be shed periodically as the animal grows

>> Bilateral symmetry (see Chapter 9 for more about symmetry)

>> Gills or gill-like organs to breathe

>> A body divided into three sections — head, thorax, and abdomen, although the head and thorax are fused into a single segment in some species

>> Several pairs of jointed appendages, such as legs and *swimmerets* (feathery paddles attached to the abdominal segments) — some appendages have evolved to develop pincers at the end

>> Two pairs of antennae

>> A nerve cord and brain

Most crustaceans are either male or female, and they reproduce sexually, but a few species, including barnacles, are *hermaphrodites* (each one can serve in either a female or male role). The female lays the fertilized eggs and either leaves them or carries them with her till they hatch. The hatchlings typically pass through a few larval stages on their way to adulthood starting with the *nauplius* stage (three pairs of appendages, a median eye, and little to no apparent segmentation).

The Shrimpy Crustaceans: Branchiopoda

Branchiopods are predominantly freshwater species, so they are outside the scope of this book with the exception of *Cladocera* (water fleas).

Water fleas make us itchy just thinking about them, but fortunately these "fleas" don't bite. As their name suggests, these creatures resemble tiny fleas, with an oval-shaped body and two pairs of antennae; a smaller pair aid in olfaction (smelling things), and a large, feathery pair help with locomotion (see Figure 11-1). The rapid movement of the antennae is also thought to contribute to the flealike behavior of water fleas — they jump around like their land-based namesakes. Another characteristic of water fleas is that they have a compound eye, which is thought to be the result of two eyes fusing together over time.

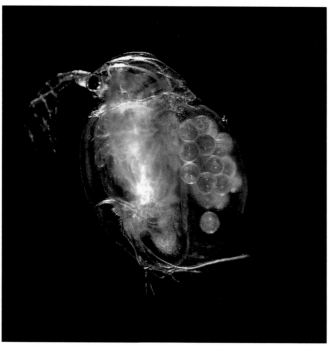

Photo by Hajime Watanabe. Licensed under CC BY 2.0.

FIGURE 11-1:
A female water flea carrying eggs.

While water fleas include only a few marine species, they can be found worldwide in estuaries, coastal waters, and the open ocean. They're most commonly filter feeders, but can also be grazers, scavengers, or predators, feeding on organic matter, phytoplankton, bacteria, or other microorganisms.

Although they're small, they're usually bigger than fleas, many reaching a size of 6 millimeters (about a quarter inch) and some being as large as 18 millimeters (about three-quarters of a inch). They're often pale or translucent, but they can take on the color of whatever they're eating, some going yellow-green after eating algae and others going pink after eating bacteria.

FUN FACT

Water fleas often reside in vernal pools, which have daily wet and dry phases. During the wet phase, water fleas may be all female, and they can reproduce all by themselves. Each female produces fertile eggs she carries with her until a wet phase when they hatch. The size of their offspring may be influenced by the presence of different predators, and in response to predators, some grow teeth on their "necks" for defense. Isn't nature awesome?!

Real Softies: Malacostraca (Soft-Shell Crustaceans)

When you think of crustaceans, you're probably thinking of something within the class *Malacostraca* — the "soft-shelled" crustaceans. Malacostraca represents about 30,000 species, including the mantis shrimp, krill, crabs, lobsters, and crayfish, which live all around the world in freshwater, marine, and even terrestrial environments. These 30,000 species can be divided into three subclasses — phyllocarida, hoplocarida, and eumalacostraca.

FUN FACT

If you ever tried to eat a whole crab or lobster, you may wonder where this nonsense about *soft-shelled* crustaceans came from. When these creatures grow, their shell doesn't, so they have to shed their shell to grow, then a new shell hardens around them. For several hours or days after shedding their shell (molting), their soft body is exposed until a new shell grows around them. Hence, soft-shelled.

Phyllocarida

Phyllocarida has three orders, two of which are extinct. The only remaining order is *Leptostraca,* sometimes referred to as leaflike shrimp, mud shrimp, or sea fleas. They look very similar to water fleas covered earlier in this chapter. They have a round body; a flattened bivalved *carapace* (two halves fused together, no hinge), which encloses the head and thorax and, in some species, part of the abdomen; feathery legs that point backward; and a long tapered tail (see Figure 11-2). They generally hang out on the seafloor from the shallows to the deep sea.

FIGURE 11-2:
Leptostraca.

Hoplocarida

Like Phyllocarida discussed in the previous section, *Hoplocarida* contains only one surviving order, *Stomatopoda* — the mantis shrimp, one of our favorite crustaceans. What's not to like? They can be quite colorful, and they generally have large inward-folding claws (like a praying mantis) with either calcified clubs or sharp tips at the ends to knock out or spear their prey (see Figure 11-3). Those equipped with clubs can deliver punches with such force that some are known colloquially as "thumb splitters." They can strike at the acceleration of a .20 cliber rifle bullet, 50 times faster than a human eye can blink, and are difficult to keep in aquariums, because a single flick of the "wrist" can shatter the glass. Their blows even generate tiny vacuums called *cavitation bubbles,* which can generate intense heat and light as they collapse. You really have to see it to believe it. Take a break and watch a YouTube video of a mantis shrimp in action. These guys are brutal. We'll still be here when you come back.

Okay, now that you saw that, you can join us in gratitude that they only grow to about 8 or 10 centimeters (3 or 4 inches) in length with the largest growing to a little over 30 centimeters (1 foot). Imagine a mantis shrimp the size of say . . . a German Shepherd! That would be terrifying.

Source: Prof Jamie Seymour – James Cook University

FIGURE 11-3:
A mantis
shrimp.

Mantis shrimp also have stalked eyes that are truly extraordinary and can move and perceive depth independently of one another. These eyes are equipped with between 12 and 16 color receptors (compared to us humans who only have 3!). We're not sure why they have so many, because they don't seem to be able to dif-ferentiate colors very well, but the increased visual data may help them navigate or hunt more effectively or make better decisions, like where to throw their next punch. They use these eyes to see a broader spectrum of light — from infrared light (which humans can't see), through visible light (which humans can see), all the way to ultraviolet light (which humans also can't see), as well as polarized light, specifically circularly polarized light, which *no* other animal can see (that we know of). They may use this polarized light to communicate with one another using signals no other animal can detect, but no one knows for sure.

Mantis shrimp are generally solitary animals, many digging their own burrows in soft sediments or hiding in crevices, depending on whether they're bashers or pokers. In either case, they're some of the most impressive and menacing crea-tures (for their size) on the planet.

Eumalacostraca

Eumalacostraca is a subclass of nearly all malacostracans that includes about 40,000 species characterized by 19 segments, a flexible abdomen, and a fan tail (which looks a lot different on a lobster than on a crab). This subclass has 16 distinct orders, which would require an entire chapter to cover. Here, we present four of the more common extant (not extinct) orders of mostly marine species.

Isopoda

Isopoda is a large order of about 10,000 species. About half of these species live on land — you probably know them as pillbugs, woodlice, or roly-polies — but they also live in freshwater and marine environments. Marine species, which account for nearly 4,500 species, typically live at all depths where they crawl along the seafloor, although some can swim short distances. They're predominantly scavengers, but can also be grazers, predators, and parasites (see Figure 11-4). Isopods are often abundant in deep-sea ecosystems, with some such as the giant isopod reaching 50 centimeters (20 inches) in length (about as long as your forearm). They're very diverse, but most of them have rigid, overlapping segments that make them look like giant cockroaches or armadillos.

FIGURE 11-4:
A giant isopod.

Source: Woods Hole Oceanographic Institution, P. Caiger – www.whoi.edu

Amphipoda

Amphipods are shrimplike creatures with laterally compressed bodies that live in both freshwater and marine environments, though the large majority of species are marine. They were even spotted at Challenger Deep (the deepest part of the Mariana Trench). Some species, such as sand hoppers (often referred to as "beach fleas"), prefer sandy beach environments. Many amphipods are burrowers, some live in tubes they build themselves, and others are *planktonic* (drifters). Their back three pairs of legs point backward and are used for locomotion — swimming, digging, or "leaping." They're grazers, omnivores, predators of small animals, or even parasites, such as whale lice (ewwww). Regardless of how creepy looking they may be, they're important components of the food web, as prey for fish, penguins, birds, and pinnipeds (walrus, seals and sea lions).

Euphausiacea

Krill! Finally! These crustaceans look a lot like shrimp, except krill are generally smaller and have three sections — head, thorax, and abdomen — whereas shrimp have only two sections — a *cephalothorax* (where the head and thorax are fused) and abdomen. Most types of krill are 1 to 2 centimeters long, about the length of one or two paperclips, but some krill can grow as long as 12 centimeters (about 5 inches). They reside in all oceans, including the polar regions, and are usually *epipelagic* (surface water dwellers). Sometimes, they form huge, dense swarms hundreds of meters across containing trillions of individuals — swarms that are so large they can occasionally be seen from space.

FUN FACT

Most species of krill are bioluminescent, producing light in a specialized organ called a *photophore*. Scientists aren't sure, but the light may serve as a signal to attract individuals to form swarms (the krill rendition of a flash mob).

Krill play a key role in many marine food webs by feeding mostly on phytoplankton and in turn feeding larger animals, such as fish, birds, squid, manta rays, and whales (see Figure 11-5). A single species of krill, the Antarctic krill, represents an estimated biomass of 125 million to 6 billion tons! That may seem like a lot, but consider that one single migratory adult blue whale needs to eat about 4 million krill (3 tons) daily, give or take a krill, to survive.

Krill also serve as a valuable carbon sink, feeding on carbon-rich algae and then excreting the carbon-rich waste, which eventually settles to the seafloor where it's stored. By one estimate, this carbon sequestration process is equivalent to removing 35 million cars from our roads, annually.

Unlike other types of plankton, which generally have short life spans, some krill can live as long as ten years. To avoid being eaten (by nearly everything), krill participate in *diel* (daily) vertical migrations, diving deep to avoid predators during the day, and coming up to feed at night.

Unfortunately, scientists believe that the numbers of krill are dropping dramatically. For example, populations of Antarctic krill are estimated to have declined by *80 percent* since the 1970s, primarily due to climate change and decreasing ice cover. The reduced ice means that the krill's main food source, ice algae, is not as available. Another factor? The growing market for krill as a supplement for people and pets is also causing widespread concern. Careful management is underway to attempt to prevent overfishing and the collapse of krill populations which is a good thing because, to put it simply, krill are the fuel that drives the engine of the planet's marine ecosystem.

FIGURE 11-5:
Krill with phytoplankton clearly visibly in its stomach.

REMEMBER

As climate change puts more stress on ocean ecosystems, protecting the environments in which krill thrive is vitally important for all ocean life, as well as our own. Visit antarctica2020.org for more information on their campaign to protect krill-friendly areas in the Southern Ocean.

Decapoda

Now for everybody's favorite softshell crustaceans — the *decapods* (literally, the ten-foots). This group includes roughly 15,000 of the most familiar crustaceans, including crabs, lobsters, prawns, and shrimp. While that's certainly a diverse cast of characters, they still have a lot in common — five pairs of "walking" limbs (one or more pairs of which are modified into pincers), head and thorax fused into a single cephalothorax and covered by a carapace, and stalked eyes, along with everything they already have in common as crustaceans.

Decapods are specialized for their environments. Hermit crabs are adapted to use the shells of other organisms for shelter and protection. Terrestrial and amphibious decapods store water to moisten their gills, so they can breathe on land. Some have flattened legs or a tail fan for swimming, or thinner, stronger legs for walking along the seafloor. Many decapods have a modified pair of legs, called *chelipeds* equipped with pincers for defense, to catch prey, or to tear food into smaller pieces to be eaten. And many are able to regenerate lost limbs.

Decapods are most commonly scavengers, but can also be filter feeders, predators, or herbivores. Most use mandibles to grind up food, and some have specialized plates in their stomachs that function like a mortar and pestle to grind their food. Members of this order are found worldwide, in most marine environments, but are more abundant in shallow waters.

SHRIMP AND PRAWNS

Shrimp have long, slender, semitransparent bodies; a flexible abdomen; a fan-shaped tail; long, whiplike antennae; featherlike appendages they use for swimming and sweeping food toward their mouths; and pincers on their front pair of legs (see Figure 11-6). They use their legs more than their tails to swim forward, but when they need to back up in a hurry, they rapidly flex their abdomen and tail. A female shrimp can lay up to 14,000 eggs, which she then typically carries on her legs until they hatch.

FIGURE 11-6:
A red night shrimp.

Source: NOAA, Public Domain

Prawns are nearly identical to shrimp but differ in several ways. Prawns are generally larger and live in freshwater or brackish environments, feeding on the bottom, whereas most shrimp live in marine environments, swimming and eating in the water column. A prawn's gills are fanlike compared to the shrimp's platelike gills. Prawns have pincers on their front three pairs of legs; shrimp have pincers only on the front pair. And the prawn's shell segments overlap down their abdomen, so the bend in their bodies is less pronounced than it is in shrimp.

FUN FACT

Several shrimp species commonly known as "cleaner shrimp" crawl around on fish, eels, and other larger marine animals to pick parasites off their bodies, gills, and mouths, keeping them healthy. It's sort of like a drive-through car wash. The shrimp and the animal being cleaned have developed ways of communicating to one another so everyone involved in the process is a willing participant and nobody will be injured (intentionally, anyway).

LOBSTERS AND CRAYFISH

Lobsters and crayfish have elongated bodies, long antennae, large pincers, and a fan tail (see Figure 11-7). The big difference between the two is that crayfish live in fresh water, whereas lobsters live in salt water. Lobsters are also a whole lot larger.

FIGURE 11-7:
An American
lobster.

Source: Keith Ellenbogen – www.keithellenbogen.com

Some crustaceans are lobsters in name only; they're not "true lobsters." For example, spiny lobsters (also known as rock lobsters) lack the big front pincers so characteristic of lobsters. Squat lobsters look like a blend between a crab and a lobster and are more closely related to hermit crabs. Slipper lobsters look like what would happen if a lobster started pumping iron — their bodies are much thicker, their legs and claws are more robust, and their antennae are shaped like paddles.

HERMIT CRABS

Hermit crabs are famous for carrying their homes on their backs (see Figure 11-8). They have a tail that's perfectly adapted to hook inside a snail shell, secure them in place, and pull them inside at the first sign of danger. As they outgrow their house, they find a more spacious one, exit the old home, and make themselves comfortable in their new abode.

FIGURE 11-8:
A hermit crab.

Source: NOAA, Public Domain

Some hermit crab relatives, such as the coconut crab, don't need a shell for protection, while others use empty polychaete worm tubes, corals, or sponges for shelter and protection. While many hermit crabs are able to leave the water for short periods of time, they must return to water periodically to wet their gills, in order to breathe.

CRABS

Crabs look like what would happen if you attached legs, pincers, and eyes to a clam. What makes them unique among decapods is their broad carapace (dorsal shield). They're also known for their large front pincers and their propensity for walking sideways, though some crabs have paddles on their back legs to help them swim. While they use their pincers mostly to eat, dig burrows, and defend themselves, they may also use them for communication; for example, males often have larger pincers than females or one pincer larger than the other and will gesture with their claws (a crab's way of saying, "come hither").

Most crabs are scavengers, playing an important role in keeping the seafloor healthy and tidy and recycling nutrients, but some are active predators. For example, the coconut crab (primarily a land crab), shown in Figure 11-9, is known to eat large birds. (Now you know why this bad boy doesn't need to wear a shell for protection.)

FIGURE 11-9: Philippe holding a coconut crab.

Source: Ashlan and Philippe Cousteau

A New Twist on Putting Your Foot in Your Mouth: Maxillopoda

Maxillopoda comes from the words "maxilla" (jaw bone) and "poda" (foot). They're characterized by a short body, reduced abdomen, and general lack of appendages (although that's not the case with all maxillopods). They account for nearly 14,000 species, which include copepods, barnacles, and fish lice. While barnacles use their legs to bring food to their mouths, most maxillopods use their *maxillae* (a pair of externalized mouth parts).

Copepods

If you're a fan of *SpongeBob SquarePants*, you're already somewhat familiar with copepods thanks to the character Plankton, who just happens to be one. Characteristic of copepods are their oval or tear-shaped bodies and long, thin antennae (see Figure 11-10).

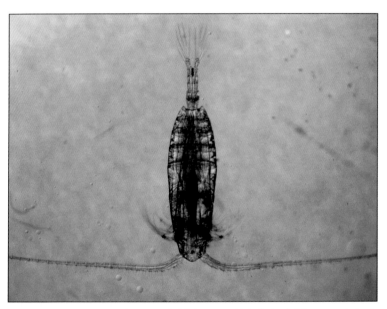

FIGURE 11-10:
A copepod.

Source: Woods Hole Oceanographic Institution, P. Caiger – www.whoi.edu

However, copepods represent a diverse collection of an estimated 12,500 species, each with unique body types suited to their particular environment. They're an important component of both epipelagic and benthic communities, they're thought by some to outnumber insects three to one, and they dominate holo-plankton communities in both numbers and biomass. (*Holoplankton* are those that spend their entire lives adrift.)

The planktonic varieties feed mostly on algae or organic matter, but some can feed on other microorganisms and zooplankton, while larger species are often para-sites of fish and invertebrates. Similar to krill, they play a key role in food webs, serving as a path of energy transfer from phytoplankton to larger animals.

FUN FACT

Copepods have a multistage life cycle encompassing as many as 12 different phases.

Barnacles

At first glance, barnacles look and act more like oysters or clams than crustaceans. The spend their youth floating or swimming in the water column before attaching themselves to a solid surface and building a cone-shaped home they live in for the rest of their lives. They'll attach themselves to nearly anything — rocks, shells, corals, turtles, whales, manatees, the underside of a boat, the piles supporting a bridge or pier. Unlike other crustaceans that use their legs mostly to walk and swim, barnacles have adapted feathery legs called *cirri* that enable them to reach out of their "concrete" bunkers and snatch particles of food.

Barnacles come in two main types — goose barnacles and acorn barnacles. Goose barnacles have a long, fleshy stalk called a *peduncle* topped with a white shell that contains the main body of the barnacle. Acorn barnacles are shaped more like short, stout mini volcanoes. Other types of barnacles exist that don't look any-thing like goose or acorn barnacles, some of which are parasites that live on or in other crustaceans, echinoderms, or cnidarians.

Fish lice

Branchiurans (fish lice) comprise roughly 150 to 200 species of small, flat, disk-shaped ectoparasites (*ecto* meaning they live on the outside of their host's body). They have mouthparts specially adapted to a parasitic lifestyle, as well as spines or suckers to help them attach to their slippery hosts (fish). They feed on the mucus, epidermal tissue ("skin"), and blood of their hosts. After they've eaten, they can last two to three weeks without feeding. Many fish lice attach behind the gills of their host, where they're slightly more protected.

A few fish lice aren't likely to permanently harm or kill the fish, but they can pose a threat when they infest fish farms.

Crustacean Cave Dwellers: The Remipedia

Remipedia is a small group of tiny crustaceans that look more like centipedes than crabs. They live in *anchialine caves* (completely dark underground caves with an opening to the ocean), and as a result they're pale and have no sense of sight (who needs sight when there is no light anyway). These cave systems are unique environments, with freshwater inputs coming from land based sources above and salt water seeping in from the coastal opening below. Remipedes generally live in the marine (salt water) sections of these caves, where availability of nutrients and sometimes even oxygen is typically low.

Like other crustaceans, they have two sets of antennae and numerous pairs of legs (far more than other crustaceans), which function more like paddles than legs. They range in size from about 9 to 45 millimeters or about a third of an inch to just under 2 inches in length. Little is known about this relatively small group of crustaceans (about 24 species). Some seem to be filter feeders, but they've also been observed grabbing other small organisms, so some may be predatory, too. Some are also venomous, making them potentially the only venomous crustaceans.

Ecologists are taking steps to preserve remipedes, fearing they might disappear before they've had a chance to be studied. Some of the cave systems they live in are being listed for protection, and better management strategies of exhaled gas are being implemented by cave divers so as not to upset the delicate balance of gases within the caves.

Ostracoda

Ostracods (commonly known as seed shrimp or mussel shrimp) are small crustaceans (ranging in size from 0.1 millimeters to 32 millimeters [.004 inches to 1.3 inches]), characterized by their bivalve-like shell, similar to that of a mussel. In open ocean environments, the shell may be smooth and transparent, while in benthic environments, it may be thicker and ornamented with spines or bumps. The shells are usually bean-shaped, but can also be oval, rectangular, or triangular, and can be different shades of brown, cream, green, red, or black and either uniformly colored or in patches or stripes. Like many other small crustaceans, they use their second antennae to help them swim or to move through sediment.

Ostracods consist of roughly 13,000 living species, along with about 50,000 to 60,000 additional prehistoric species. Fossils of ostracods have been found as far back as the Cambrian period (see Chapter 3). Today, they live in both marine and freshwater environments, from temporary pools to warm waters to the polar regions to the deep sea and even on land, with some species living in the soil in the humid tropics. They have a wide range of diets, but usually feed on microbes and organic material.

Some species, such as the sea-firefly (known in Japan as *umi-hotaru*), are bioluminescent, producing a brilliant blue glow. Interestingly, for some ostracods, it's not the individuals themselves that are glowing — it's their vomit (beautiful). They expel a small amount of mucus mixed with some enzymes and reactants, which produce the light. Males are thought to use this for courtship displays! Each species may even have its own specific light display pattern. Baby, you had me at barf.

What about Horseshoe Crabs?

Sorry, but horseshoe crabs didn't make the cut because (surprise!) they're not crabs. They're not even crustaceans. They're more closely related to spiders and scorpions. So, what are they doing in this chapter? We included them here because, well, they're called crabs, and we don't want you to think that we omitted them by mistake. Also, if you're ever on a beach where these weird-looking creatures gather to mate in the spring, you'll be wondering, "What the heck are these things?" And if you try looking it up in this book, we want you to find the answer.

Horseshoe crabs are considered living fossils. They've been around for nearly 445 million years and have changed little, if any. From above, the horseshoe crab looks like a military tank with a tail (see Figure 11-11). It has a huge carapace that completely covers its body (mouth, legs, gills, stomach) except for its tail and its ten "eyes" — a pair of fully functional eyes on top, and eight photoreceptors in other locations, including the tail. Horseshoe crabs are sort of like robotic vacuum cleaners, crawling across the seafloor consuming anything edible that suits their taste.

Horseshoe crabs take about 9 to 11 years to reach sexual maturity, at which time they stop molting and begin participating in annual spring migrations to shoreline spawning areas. (All the horseshoe crabs you see on a beach in the spring are mature.) Few hatchlings make it through the natural predator cycle, but if they survive this and other rigors, they can live for almost 20 years.

FUN FACT

Horseshoe crab blood is a valuable commodity that's used to test the purity of medications. This blood is particularly sensitive to bacteria, and when a horseshoe crab is wounded, the blood cells quickly swarm to the wound, form a clot, and destroy the invading bacteria. Scientists created an extract called Limulus

Amebocyte Lysate (LAL) from this blood, which they use to test medicines for the presence of biotoxins that can be fatal to humans. Recently, horseshoe crabs have made the news for playing a key role in testing potential COVID-19 vaccines. Another fun fact is that horseshoe crabs are genuinely blue bloods — unlike our blood, which is high in iron, theirs is high in copper, making it *blue*.

FIGURE 11-11:
A horseshoe crab.

Source: Keith Ellenbogen – www.keithellenbogen.com

Chapter **12**

Getting Chummy with Fish: Bony and Otherwise

W hy do fish live in salt water? Because pepper makes them sneeze. (Sorry, we couldn't resist.)

The ocean is swimming with all kinds of fish — tuna, sardines, anchovies, barracuda, grouper, flounder, sharks and rays, puffers, sailfish, swordfish, snapper, stingrays, bass, tarpon, marlin, hogfish, hagfish, and the list goes on and on. Holy mackerel!

What do they all have in common? What makes a fish a fish? Actually, *all* fish have only a few obvious characteristics in common — an encased brain (skull) and an obvious head with eyes, teeth, and sensory organs (and no, that doesn't make you a fish). *Most* fish are *vertebrates* (they have a backbone), live in water, breathe primarily with gills, have paired limbs in the form of fins, are covered in scales, and are *cold-blooded*, meaning they can't control their own body temperature. But of course some fish have to be different. Tuna are warm-blooded, lampreys don't have paired fins, mudskippers and lungfish can breathe air, and many species don't have scales.

According to at least one estimate, the ocean is home to more than 30,000 species of fish, and we can't possibly cover all of them in one chapter. We can't even cover all our favorites in one chapter. So, to make this topic more manageable, we've decided to school you on the three main classes of fish — jawless, cartilaginous, and bony –- and introduce you to a few of the more popular and fascinating species and individuals within each group.

Look Ma, No Jaw! Agnatha

Imagine a creature with the body of an eel and the mouth of a leech, and you have a pretty good idea what an agnathan looks like — a jawless fish. They're also missing a stomach and paired fins, but they have a light-sensitive "third eye" called a *pineal eye,* which is essentially a simple light receptor. They're shaped more like eels or snakes and have a cartilaginous skeleton like sharks (see the next section for more about cartilaginous fish). They also have gill pores as opposed to slits and possess a *notochord* — a flexible, rod-shaped supporting structure that runs the length of their body and is an evolutionary precursor to a vertebra.

Members of the superclass *Agnatha* (Greek for "no jaws" and pronounced *ag*-nuh-thuh) go way back and may be the earliest vertebrates ever. According to fossil records, they've been around for 450 million years. Now the ocean is home to two types of agnathans — lampreys and hagfish.

Lampreys

Lampreys consist of about 40 species that live in marine and freshwater environments. They range in length from about 15 to 100 centimeters (6 to 40 inches). They have one or two *dorsal* (top) fins, a *caudal* (tail) fin, two eyes, a single nostril, and seven gill openings on each side of their body.

Most lampreys (all marine species of lampreys) are parasites. Their mouths look like a suction cup with tiny teeth (see Figure 12-1), which they use to scrape a hole in the host through which they suck the host's bodily fluids. Adults spawn and lay eggs in fresh water, and the hatchlings live as larvae (called *ammocoetes*) for one or more years buried in the silt and feeding on microorganisms before metamorphosing into adults.

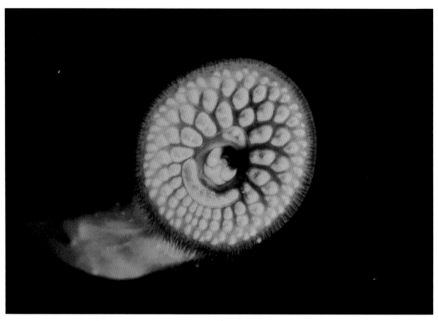

Source: U.S. Fish and Wildlife, Public Domain

FIGURE 12-1:
A lamprey's
mouth.

Hagfish

Hagfish, also known as "slime eels," aren't exactly eels, but they sure look like eels, with their tubular bodies and paddle-shaped tails, and boy do they produce a lot of slime, like buckets of it. This isn't your average, ordinary icky fish mucus slime; it's more like rubber cement slime. A hagfish uses it for self-defense, releasing loads of it when attacked, clogging the mouth and gills of its assailant (take that!). We can't imagine a predator who wouldn't gag trying to eat one of these slimy, grayish pink, eel-looking things.

Two other unique characteristics of hagfish are that they're the only known vertebrates without a backbone, and they're covered in very loose-fitting skin. These two qualities, along with their ability to produce loads of slime, make them both an effective predator and challenging prey. As a predator, the consistency of the hagfish's skin and body enable it to squeeze into tight crevices to pursue its food. As prey, the hagfish's loose skin and squishy body coupled with its ability to release its slime when attacked means it can literally tie itself in knots and slither away, leaving its attacker with a mouthful of goo.

Unlike lampreys, which have a keen sense of sight, hagfish are nearly blind, relying on smell and touch to find their prey — usually worms and other small invertebrates. Hagfish are also known to slither inside dead or dying animals, where they consume nutrients not only through their mouths but also through their semipermeable skin, sort of like how our intestines extract nutrients from food.

Look Pa, No Bones! Chondrichthyes

Chondrichthyes (pronounced kon-*drick*-thees) is a class of fish with skeletons made mostly of hard cartilage instead of bones and include everyone's favorite — sharks and rays. But that's not all they have in common. Most chondrichthyans are dark (usually gray) on top and white on the bottom — a coloration scheme called *countershading* that makes them blend in with the brightness of the sky shining from above when viewed from below, and blend in with the darkness of the depths beneath them when viewed from above looking down. They generally have five gill slits, typically located just behind the head (but on the undersides of skates and rays). Most species lack a swim bladder like other fish have to control their buoyancy, but almost all chondrichthyans are pretty much neutrally buoyant or slightly negatively buoyant due to the oils stored in their livers.

Chondrichthyan skin is covered with toothlike scales, called *placoid scales*, which take the form of spines in some species, such as the rays. Placoid scales (also called *dermal denticles*) are pointed like tiny teeth, with the points all facing the same direction toward the tail. So if you pet a chondrichthyan from head to tail, the skin feels smooth, but rub your hand from tail to head and the skin feels like sandpaper. Unlike the scales on most fish, placoid scales do not become larger as the animal grows.

Like most vertebrates, chondrichthyans have the standard five senses — sight, smell, touch, hearing, and taste. Because their sense of sight isn't always the greatest, they have two additional sensory systems to help them compensate:

>> **The lateral line system of navigation:** Along the left and right side of the fish, buried under the skin, is a lateral line or canal that contains a series of sensory organs. Tiny tubes lead from this canal to pores on the surface of the skin. This system enables the fish to pick up on vibrations in the water and changes in water pressure caused by the fish's movement in relation to objects and other fish.

>> **The ampullae of Lorenzini:** Sharks and rays have electrical receptors located near their snouts that open to facial pores. Whenever a fish or other animal moves in close proximity to the face of a shark or ray, the shark or ray can detect the very weak electrical signals emitted by the animal's muscle movements (even their heartbeat). This is especially useful when a shark attacks, because they typically roll their eyes back or close them tight to protect them during mealtime.

Unlike bony fish such as tuna and salmon that spawn and lay eggs, chondrichthyans rely on internal fertilization. Males have reproductive appendages called *claspers* near the inner margins of their pelvic fins. To mate with a female (and keep from drifting apart while they do), the male inserts one of his claspers into the female and inseminates her. Fertilization of eggs is always internal. Then, depending on the species, the mother bears a litter of offspring and either lays the fertilized eggs or carries them with her until they hatch and are birthed live.

In the following sections, we look at the two subclasses of chondrichthyans — Elasmobranchii (sharks, rays, and skates) and Holocephali (chimaeras).

Elasmobranchii: The fish with a PR problem

Elasmobranchii (pronounced ee-laz-mo-*brank*-ee-eye) is a familiar subclass of fish that includes some of the most misunderstood and maligned creatures in the world — sharks, skates, and rays. Elasmobranchs are characterized by a rigid dorsal fin (the top fin) and have four to seven pairs of gill slits to breathe. They're all carnivorous, feeding on everything from large marine mammals to small crustaceans, and they lack a swim bladder, equipped instead with an oily liver to maintain buoyancy. In contrast to the teeth of most vertebrates, which are locked into sockets in the jaw bones, elasmobranch teeth are attached to the jaw with fleshy tissue. Many species have rows of teeth that continue to be replaced — some may go through as many as tens of thousands of teeth in a lifetime.

Selachii: Sharks

You may not be the biggest shark fan, but *I* (Ashlan) am a HUGE shark fan. I'll tell you why in a minute, but first, we want you to get to know a little about these amazing fish.

Sharks go way back. The first evidence of sharks dates to the Ordovician period (see Chapter 3) 400 to 450 million years ago! Approximately 450 species of fish have the honor of being classified as sharks, and they range in size from about 20 centimeters (8 inches) to 12 meters (40 feet) long. They're all predators and feed mostly on fish, seals, and whales, but some sharks, such as the whale sharks and megamouth sharks, feed on tiny plankton.

FUN FACT

Baby sharks are called "pups," but even before they're born, they're not all sweet and cuddly. Some baby sharks eat their brothers and sisters who are growing slower while still inside their mother (talk about sibling rivalry!). Other species of sharks create extra eggs for the growing pups to consume. Both are examples of *intrauterine cannibalism*. Some scientists believe this practice is a case of "survival of the fittest" and helps keep the species strong.

In this section, we cover a few of our favorite sharks.

The smallest of sharks is the dwarf lantern shark. Lantern sharks, as their name implies, are *bioluminescent* — able to produce their own light, which they use to ward off predators, blend in with the lighter background above them, and communicate when swimming in schools.

On the other end of the size spectrum is the whale shark (see Figure 12-2), which can grow as long as 12 meters (40 feet) and weigh up to 40 tons. While in theory a whale shark could easily gulp down prey larger than humans, this extraordinary creature only feeds on tiny plankton. So, the biggest fish in the sea eats some of the smallest food.

FIGURE 12-2:
The largest fish in the sea, the whale shark, followed by a diver about to attach a satellite tag to monitor the animal's movements for research.

Source: Michael Muller – www.mullerphoto.com

Bull sharks (see Figure 12-3), so named because of their short, blunt snout, wide body, and aggressive temperament (and because they often head-butt their prey before eating them), are unique in that they can live in freshwater or marine environments, sometimes swimming far upstream into a river or tributary. Among sharks, they pose perhaps the most serious threat to humans mostly because they like to swim in the same places humans do, not because they find humans particularly tasty.

Mako sharks (see Figure 12-4) are perhaps the fastest of the species and one of the fastest fish on the planet, attaining speeds of up to 74 kilometers (45 miles) per hour, fast enough to chase down its favorite food — the speedy tuna.

FIGURE 12-3:
The bull
shark —
salt water,
fresh water, no
problem. (This
photo is from
our friend
Michael Muller,
who adds a
little artistic
flair to his
photographs,
which helps
illustrate the
beauty and
drama of
these fantastic
creatures.)

Source: Michael Muller – www.mullerphoto.com

FIGURE 12-4:
The mako
shark is
known for its
speed and
athleticism.

Source: Michael Muller – www.mullerphoto.com

The hammerhead shark gets the award for being the weirdest looking with an eye on either end of its mallet-shaped head. They feed mostly on small fish, octopus, squid, and crustaceans and can grow up to 6 meters (20 feet) long and weigh as much as 450 kilograms (about 1,000 pounds). Depending on the species, they are either solitary (see Figure 12-5) or school in enormous numbers (see Figure 12-6).

FIGURE 12-5:
A lone great
hammerhead
shark.

Source: Cesere Brothers – www.ceserebrothers.com

FIGURE 12-6:
Schooling
hammerheads.

Source: Michael Muller – www.mullerphoto.com

No section on sharks would be complete without mention of the baddest shark prowling the oceans — the great white shark (see Figure 12-7), which can grow up to 6 meters (20 feet) long and weigh more than 2,268 kilograms (5,000 pounds). They have a varied diet of fish, crustaceans, seals, sea lions, other sharks, and even small-toothed whales such as orcas. Where do they live? Wherever they want. But seriously, you can bump into one just about anywhere the water temperature is between 12 and 24 degrees Celsius (54 and 75 degrees Fahrenheit).

FIGURE 12-7:
The great
white shark.

Source: Michael Muller – www.mullerphoto.com

Some people are terrified of sharks and won't even take a dip in the ocean because of them, but we love sharks. I (Ashlan) am particularly fond of them for their beauty, size, power, athleticism, diversity, and for all they do to keep our oceans healthy. We really want you to love and respect them, too . . . and, if you fear them, please stop being afraid. Sharks have far more reason to fear (and hate) us humans than we have to fear them.

Yes, *Jaws* scared the "carp" out of all of us, and to his dying day Peter Benchley (who wrote the book) felt horrible for this. He later became a huge shark activist, but the damage had been done. Generations of readers and moviegoers were scared out of their swimsuits of sharks, especially the great white, and unnecessarily so.

Sharks are not vicious murderers just waiting for you to wade past the buoys or paddle your surfboard overhead. Certainly, some animals kill for fun — namely dolphins, house cats, killer whales, leopards, honey badgers, and, of course, humans. But the vast majority of predators, including sharks, eat only when necessary. And, like many predators, sharks carefully calculate their return on investment — whether they'll expend more energy attacking potential prey than they'll receive from eating it. That's why sharks smell for blood. They're on the prowl for the wounded and the sick, and they can go long stretches without eating, so they can afford to be picky eaters and wait for the right opportunity to come along. Nothing personal, but you're not the first choice on their menu. In fact, scientists believe that shark attacks happen as a case of mistaken identity. Sharks can only tell what you are with their mouth (no hands to feel ya with) which is why most attacks are a single bite and release and thus not usually fatal. The sharks quickly realize that you aren't their normal prey and they move on.

Still afraid? Then check out Table 12-1 to put your chances of dying from a shark attack in perspective.

TABLE 12-1   ## Gauging the Risk of Death from Shark Attack

Cause of death	Average annual deaths in the U.S.
Car accident	44,757
Gun deaths	36,000
Accidental poisoning	19,456
Falling	17,229
Bike accident	762
Air/space accident	742
Excessive cold	620
Sun/heat exposure	273
Bee, wasp, hornet stings	62
Lightning	47
Train accident	24
Dog attack	16
Fireworks	11
Spider bite	7
Snake bite	5
Shark attack	1

Not only are sharks much less dangerous than many people think, they also play a vital role in maintaining a healthy ocean. As apex predators, they keep the population of their prey in check, strengthen the gene pools of their prey, and reduce the spread of disease by eating the sick, weak, and injured. Lions, tigers, bears, and other apex predators perform the same service, but they don't get the horrible rap that sharks do (and they also attack people every once in a while, just sayin'). Sharks also protect plants and help preserve plant-based ecosystems by reducing the populations of the animals that graze on those plants. And they do their part to sequester carbon.

Philippe and I swim with all types of sharks all over the world. From dozens of great whites off Mexico, swarms of grey reef sharks in the Marshall Islands, to whale sharks in La Paz and huge great hammerheads in the Bahamas — never once have we felt scared or threatened. But we are always cautious and respectful when we're in the water with these extraordinary predators. The ocean is their home, not ours after all.

Unfortunately, every year, sharks are killed for their fins and their meat — about 100 million a year (but that number could actually be anywhere between 63 million and 273 million sharks killed each year). On average, that means about 11,400 sharks are slaughtered every hour, whereas four people in the whole entire world die from shark attacks per year on average. Sharks have far more to fear from us than we do from them.

Batoidea: Rays

Rays are a group of about 500 species that live in ocean waters worldwide. They look like a shark that has been smushed into a pancake in the outline of a kite (see Figure 12-8). While sharks propel themselves with their tail, rays propel themselves with elongated winglike pectoral fins. Some rays have a whiplike tail tipped with a venomous barb, which they use for self defense. A ray's mouth is usually on the underside of its body and, when viewed from certain angles, looks as though it's smiling. Instead of pointy teeth like sharks, rays have evolved rounded teeth they use to crush and grind their prey, mostly mollusks, crustaceans, and small fish.

Source: Cesere Brothers – www.ceserebrothers.com

FIGURE 12-8:
A group of stingrays.

Another difference from sharks is that rays have their gill slits underneath their body as opposed to the sides and take in water to breathe through large openings (called *spiracles*) on the upper surface of the head. In addition, most rays have their eyes on top of their head. Most are *benthic* (bottom dwelling), but some species, such as the manta ray, are *epipelagic* (free swimming). Like sharks, males have a clasper they use to mate with the female, which then almost exclusively gives birth to live young. Finally, rays are almost all marine.

Rays are a diverse group that includes electric rays, stingrays, manta rays, and more. In this section, we cover a few of our favorites.

Electric rays have a rounded body and range in length from less than 30 centimeters (1 foot) to about 2 meters (6 feet). They're equipped with two large electric organs positioned on either side of their head that are capable of delivering a 220-volt shock — more than sufficient for stunning prey and fending off predators. They generally feed on small fish and invertebrates, so you're pretty safe unless you happen to step on one.

The largest ray is the giant manta ray, shown in Figure 12-9, which has a "wingspan" of up to nearly 9 meters (30 feet). They're commonly referred to as "devil rays" because of the two special flaps at the front of their heads called *cephalic lobes*, but devil rays are very sweet. They feed mostly on plankton, using those devilish horns to direct more water and food into their mouths.

Source: Cesere Brothers – www.ceserebrothers.com

FIGURE 12-9:
A manta ray, sometimes called a devil ray.

Eagle rays have beautiful spotted skin that produce the appearance of dappled sunlight as they swim through the water. Unlike most rays, they have a very long pronounced tail and have been known to launch themselves out of the water in dramatic displays, spinning and flipping in the air.

Last on our list of rays is the most unique of the group — the sawfish, also referred to as a carpenter fish. Granted, it looks more like a shark, but its mouth and gills are positioned on its underside, and it has winglike fins characteristic of a ray. Its coolest feature is its long snout rimmed with exposed teeth.

Batoidea: Skates

Skates are members of the same subclass as rays, and they look like rays, but they have a few key differences. While rays are more diamond-shaped, a skate's body is more triangular or rounded, and they often have a pointy nose (see Figure 12-10). Skates also have thicker, wider tails with sharp spikes that run along the middle of the tail and up their backs (no barb at the tip). Skates have small teeth to eat prey, in contrast to the rounded teeth rays use to crush and grind their food. While rays swim in both shallow and open water, skates typically hang out near the bottom and often hide in the sand. Also, instead of live births, skates lay eggs in a leathery case called a *mermaid's purse*.

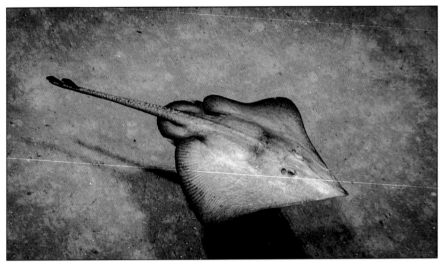

FIGURE 12-10:
A male skate.

Source: NOAA Office of Ocean Exploration and Research, Gulf of Mexico 2018, Public Domain

Unfortunately, skates are struggling to survive. The International Union for Conservation of Nature (IUCN) has listed the common skate as an endangered species in 2000 and as a critically endangered species since 2006.

Holocephali: Chimaeras

Chimaeras (also spelled *chimeras*), referred to as ghost sharks, ghost fish, rat fish, spookfish, and rabbit fish, look more like bony fish than like cartilaginous fish, but they do have pronounced pectoral and pelvic fins that move in a flapping motion, like rays and skates. They have two dorsal fins, the front of which is preceded by a sharp spine able to deliver a venomous jab; large eyes; and a slender tail that's shaped like a rat tail in some species. Unlike their cartilaginous cousins, chimaeras (Figure 12-11) have a single gill opening on each side of their body covered by a flap. The group has about 47 species that range in length from 60 to 200 centimeters (24 to 80 inches).

FIGURE 12-11:
A long-nosed
chimaera.

Source: Woods Hole Oceanographic Institution – www.whoi.edu

Some species live in rivers, estuaries, or shallow marine habitats, but many live in the deep ocean. They're often divided into three groups: the rabbit fish (with a cone shaped-nose), the elephant fish (with a hoe-shaped flexible nose), and the rhino fish (which has a long, pointed snout). Like their cousins the elasmobranchs, chimaeras have "teeth" that have fused together to create plates, one on top, and two on the bottom, that they use to crush their prey.

Check Out the Bones! Osteichthyes

Osteichthyes (pronounced ah-stee-*ick*-thees) is a class of about 28,000 fish characterized (in most forms) by a bony skeleton, scales (some without), paired fins, a single pair of gill openings supported by bony gill arches each covered by an operculum, jaws, a mouth with many teeth, a *swim bladder* (air-filled sack) for buoyancy, paired nostrils, and external fertilization of eggs.

Bony fish are generally broken down into two groups (see Figure 12-12):

>> **Ray-finned fish:** Ray-finned fish have fins supported by spines; the fins are moved by muscles in the body of the fish that do not extend into the fins. They (usually) have a pair of pectoral fins and a pair of pelvic fins, along with dorsal, anal, and caudal fins, and most look like, well, a fish — their bodies are oval-ish and tapered at each end. They breathe primarily through gills.

And they have a *lateral line* — a row of organs that sense changes in water pressure, movement, and vibrations (kind of like the sensitive hairs in our inner ears) which helps them find prey and not become prey themselves.

>> **Lobe-finned fish:** Lobe-finned fish have rounded fins (shaped more like your earlobes), which are supported by muscles and articulated bones (bones that meet to form joints). In other words, their fins are more like hands than typical fish fins, though the end of each fin has thin, bony structures that fan out from the core of the fin. These fins are thought to be the precursors of amphibian legs and feet. Lobe-finned fish also have two dorsal fins (instead of one), and they have both gills and lungs (though they're not functional in certain species), which enables some of them to breathe on land or underwater.

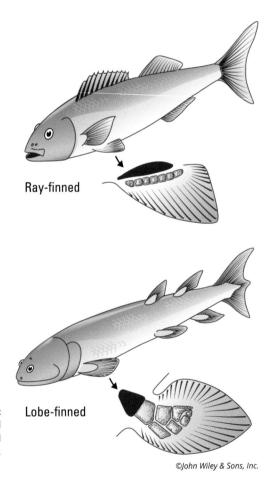

Ray-finned

Lobe-finned

FIGURE 12-12: Ray-finned and lobe-finned fish compared.

©John Wiley & Sons, Inc.

REMEMBER

As your mother told you when you experienced your first heartbreak, there's always more fish in the sea. With bony fish, that means 28,000 species and counting. We can't possibly cover them all, so we kept this section brief by describing the two groups (ray-finned and lobe finned) and highlighting the common, the cool, and the weird in each group.

Ray-finned

Actinopterygii (ray-finned fish) are, by far, the largest and most diverse group of bony fish. They include anchovies, barracuda, catfish, cod, eels, flounder, flying fish, frogfish, gars, grouper, grunt, halibut, herring, hogfish, jacks, lionfish, mackerel, marlin, minnows, mullet, needlefish, perch, piranha, pompano, porgy, sailfish, sardines, sea bass, sea trout, seahorses, shad, sheepshead, snappers, snook, sole, sturgeon, swordfish, tarpon, triggerfish, toothfish, tuna . . . we could go on but we think you get the idea.

But let's look at some notable members of the group. Imagine an award show with categories for the smallest, biggest, fastest, and more

Smallest: Stout infantfish are barely as long as the width of a pencil — females max out at about 8.4 millimeters (.33 inches) long, while males grow to only 7 millimeters (.28 inches). They're also the smallest and lightest of all known vertebrates. If you ordered a pound of stout infantfish from the menu, you'd get a plate of roughly 500,000 fish!

Largest: The sunfish (Mola Mola) can grow to be over 2,268 kilograms (5,000 pounds). They live in warm water and look like a swimming fish head that has been smushed. See Figure 12-13.

Fastest: Marlin have been clocked at a maximum speed of about 129 kilometers per hour (about 80 mph). See Figure 12-14 for a striped marlin in action.

Longest: The giant oarfish is a serpent-looking pelagic fish up to 8 meters (26 feet) long. They're silver with a red mohawk looking fin. They cherish their privacy and are rarely seen alive.

Coolest: The Antarctic icefish lacks hemoglobin in its blood, making it clear and less susceptible to freezing, sort of like antifreeze, which is useful for any fish living in the frigid waters off Antarctica. Due to this lack of hemoglobin, they have huge hearts and almost four times the amount of blood as other fish.

Weirdest couple: Female anglerfish have it going on! Living in the dark, deep ocean, these ladies have a dangling dorsal fin tipped with luminescent skin, which functions as a fishing rod protruding from the center of their head ending in a glow-in-the-dark lure just in front of their huge mouth full of teeth. And a female anglerfish has all the guys attached at the hip, seriously. Male anglerfish are usually much smaller than the females. When a male finds one of these lovely ladies, he attaches himself to her, as a permanent parasite. After a while, he loses his eyes and all his organs except his testes, physically fusing his body with hers. Talk about a Stage 4 clinger! Females can carry around multiple mates. See Figure 12-15.

FIGURE 12-13:
A sunfish.

Source: Keith Ellenbogen – www.keithellenbogen.com

FIGURE 12-14:
A striped marlin feeding on a bait ball.

Source: Pier Nirandara – www.piernirandara.com

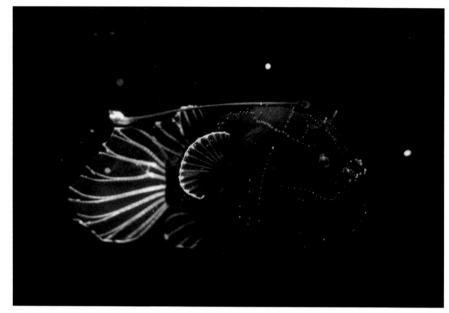

FIGURE 12-15:
A female
anglerfish —
notice its
"fishing pole"
appendage
is angled
backward.

Source: Woods Hole Oceanographic Institution, L. Madin – www.whoi.edu

Cutest: Seahorses get by on their looks (see Figure 12-16). Remember when we said most ray-finned fish look like fish? Well, seahorses are one of those exceptions. They swim around upright with what looks like a little armored body and quick-moving fins. They have a tubular snout used for eating plankton and fish larvae. And it's the male seahorse that carries around the fertilized eggs in a special pouch until they hatch — how adorable is that?!

Father of the year: Speaking of caring fathers, the jawfish is a paternal mouth-brooder, meaning he holds the fertilized eggs in his mouth till they hatch (see Figure 12-17). Every so often, he spits them out, a practice called *churning*, which keeps the eggs aerated, hydrated, and clean, thereby increasing the chances that more will hatch. All the while, he doesn't eat a thing. Fortunately for him, the incubation period for the eggs is only five to seven days.

Friendliest: Yep, fish can be friendly, and a perfect example is the grouper. Because of their large size (the goliath grouper can grow to be almost 363 kilo-grams [800 pounds]), when they reach full size, large grouper aren't really scared of anything and are naturally curious. We have had many encounters with grouper fish over the years, and honestly, they are the Golden Retrievers of the sea. The grouper in Figure 12-18 actually took Ashlan on a tour of his home, following her along her entire dive and nudging her when she wasn't paying him enough attention.

FIGURE 12-16:
Seahorses
are the cutest
critters,
and pygmy
seahorses are
the cutest of
the cute.

Source: Cesere Brothers – www.ceserebrothers.com

FIGURE 12-17:
Mr. Mom,
the jawfish,
incubating the
fertilized eggs.

Photo by Tracy Candish. Public Domain 1.0

Source: Ashlan Cousteau

Longest-living: The orange roughy, a brightly colored and slow moving fish, can live up to 200 years — that is, unless it's eaten (by us). The problem with eating orange roughy is that they don't reach sexual maturity until the age of 20, so they're *extremely* susceptible to overfishing, and they're caught using the super destructive fishing method of bottom trawling. If you need another reason not to eat them, they're often full of heavy metals, such as mercury, because they live a long time and bio-magnify toxins into their tissue. So next time you are at a restaurant and they offer you orange roughy as a special, just say no and feel free to tell the waiter (and the chef) why. Honestly, most people, even in restaurants, don't know how bad and destructive catching some of these fish is. A little friendly schooling goes a long way.

Lobe-finned

Sarcopterygii (lobe-finned fish) are a group of only eight living species with lobe-shaped fins, two dorsal fins, gills, and lungs (which aren't functional in certain species). They're broken down into two classes — coelacanths and lungfish.

Coelacanths

Thought to have gone extinct 66 million years ago with the dinos, a coelacanth was spotted by a lucky scientist at a fish market in 1938. Two *extant* (living) species are now recognized — the West Indian Ocean coelacanth and Indonesian coelacanth. Referred to as a living fossil, many scientists see them as a possible link between sea creatures and four-legged land animals, such as salamanders, newts, and lizards. Living as deep as 610 meters (2,000 feet), these carnivorous fish sleep in caves during the day and hunt for prey (mostly octopus, squid, and cuttlefish) at night. They grow to be about 1.8 meters (6 feet) long and weigh up to about 91 kilograms (200 pounds).

In 2010, Laurent Ballesta led a confidential Gombessa expedition in Jesser Canyon, Sodwana Bay, South Africa, to capture the very first pictures of the coelacanth (see Figure 12-19) taken by a diver (at 120 meters [394 feet] deep). This was possible thanks to the South African diver Peter Timm, who was the first to face the mythical fish locally named Gombessa. Discover more in the book, *Gombessa, Meeting with the Coelacanth* (`https://laurentballesta.com/`).

Coelacanths have lungs during their embryonic development that stop developing as the fish grows and are barely noticeable (and non-functioning) in adults. They also have a *rostral organ* in their snout that's part of an electro-sensory system and a hinged skull that enables the back of the skull to tilt up to enlarge the opening of the mouth.

Lungfish

As their name implies, lungfish can breathe air. Yep, they have either one or two lungs (depending on the species) to extract oxygen from the air, and they also have gills to extract oxygen from water. These are freshwater fish that live in rivers and lakes in Australia, Africa, and South America. And they're big — most species growing up to 1.2 to 2.1 meters (4 to 7 feet) long. Lungfish have horrible eyesight, but their great sense of smell, lateral lines, and sensitive touch (with their pectoral and pelvic fins) make up for it. African lungfish can also go into a deep sleep (*estivation*) for up to two years — a slimy sleeping beauty.

FIGURE 12-19: Gombessa, the local name for the South African coelacanth.

Source: Laurent Ballesta – laurentballesta.com

Chapter **13**

Meeting a Few Marine Reptiles

During the Mesozoic era, also called the Age of the Reptiles (252 to 66 million years ago), huge reptiles ruled the ocean just as their terrestrial cousins (the dinosaurs) ruled the land. Mammals were relatively small, while ichthyosaurs and mosasaurs terrorized the sea as apex predators. While the blue whale is the largest animal of all time, land or sea, growing up to about 30 meters (100 feet) and weighing more than 180,000 kilograms (400,000 pounds), back in its time, *Shastasaurus* (an ichthyosaur) was no lightweight, growing up to an estimated 26 meters (85 feet), approaching the size of a blue whale.

Since then, Earth has cooled dramatically. The average ocean temperature has dropped from about 35 to 17 degrees Celsius (95 to 62 degrees Fahrenheit), and research suggests that colder climates drive the evolution of larger mammals and smaller reptiles. Today, the largest reptile is the saltwater crocodile, which maxes out at about 6 meters (20 feet) long and 1,075 kilograms (2,370 pounds). While living reptiles (land and sea) account for more than 12,000 species and subspecies, only about 100 species live in marine environments, and, for the most part, they're relatively small compared to large sharks and whales.

All reptiles are *vertebrates* (have a backbone), are cold-blooded (mostly), have scales or *scutes* (bony plates, as in turtle shells), produce eggs (though a few give birth to live young), and have lungs to breathe. Marine reptiles have one additional adaptation — salt-secreting glands, typically located in or around their heads, for expelling excess salt from their bodies. In this chapter, we describe the four groups of marine reptiles living today and introduce you to a few representative species in each group.

Everybody's Favorite: Sea Turtles

With their armor-plated mobile homes; thick, scaly skin; and lumbering gait, turtles look prehistoric. In fact, the first land turtles evolved more than 200 million years ago, shortly after the first dinosaurs. Sea turtles followed about 50 million years later. One of the largest of the prehistoric sea turtles was the *Archelon*, which grew up to 4.5 meters (about 15 feet) with massive flippers and a voracious, carnivorous appetite. Like today's leatherback turtle, *Archelon's* shell was made of a leathery material rather than scutes.

In one form or another, sea turtles have been around a long time and today can be found throughout warm and temperate seas all over the world. They constitute a family of reptiles called *Cheloniidae*, which exhibit a few modifications that distinguish them from their terrestrial and freshwater cousins and help them survive in a marine environment. The first and most obvious is the turtle's "legs," which are more like flippers.

Another important modification is the shell structure. Like other turtles, the sea turtle has a shell made of two parts — the *carapace* (top) and *plastron* (bottom). The carapace of nearly all turtles, including sea turtles, is covered in hard plates called *scutes*. However, the leatherback turtle, as its name implies, has a leathery carapace, instead. Unlike most turtles and tortoises, sea turtles can't pull their head and legs completely inside their shell for protection, because there just isn't enough room. They've evolved a shell that's much more streamlined, facilitating better movement through water.

Like all marine reptiles, sea turtles have evolved the ability to secrete excess salt, in their case, through their tear ducts. Despite these adaptations, sea turtles are still air breathers and must return to the surface periodically for oxygen.

Meet the family

The sea turtle family is rather small, consisting of only seven species — leatherbacks, green sea turtles, hawksbills, loggerheads, flatbacks, Kemp's ridleys, and olive ridleys. In this section, we introduce you to the whole family.

FUN FACT

Sea turtles spend nearly their entire lives at sea. Only females and babies spend time on land — females to lay eggs and babies to hatch. And as soon as those babies hatch they make a mad dash to the sea.

Leatherbacks

Leatherbacks are the big mamma jammas (see Figure 13-1). An adult leatherback can reach the size of a small car — about 2.5 meters (8 feet) long weighing about 907 kilograms (2,000 pounds). Their fins make them look even larger. They earn their name from their leathery shell, and they really haven't changed much since they were swimming around with dinosaurs, but who needs evolution when you're perfect?

FIGURE 13-1:
An adult
leatherback.

Source: NOAA, Public Domain

FUN FACT

Leatherbacks are considered to be the only warm-blooded reptiles. They have a special blood vessel structure (called a *countercurrent exchanger*), which enables them to maintain their body temperature in colder waters, giving them a wider range than other sea turtles. They swim all over the world, migrating more than 10,000 miles every year, between food and nesting areas. They feed primarily on jellyfish and *salps* (sea squirts), which is pretty amazing given that these foods are 95 percent water. Also amazing is that they've been recorded to dive nearly 1,220 meters (4,000 feet) deep and can hold their breath for up to 85 minutes.

Even though leatherbacks are estimated to live as long as 50+ years and possibly to the age of over 100, the species is in serious danger. According to the National Oceanic and Atmospheric Administration (NOAA), some nesting sites have experienced a 90 percent decrease in activity in just the past three generations.

Green sea turtles

Green sea turtles (see Figure 13-2), unlike other sea turtles, are herbivores. Eating lots of algae and seagrass makes the fat under their shells green, hence their name. They live all over the world and are the largest of all the hard-shell sea turtles.

Source: Cristina Mittermeier – www.sealegacy.org

FIGURE 13-2:
A green sea turtle.

Hawksbill turtles

Hawksbill turtles are so named because of their beaklike mouths, which they use to eat sea sponges, small fish, algae, and more (see Figure 13-3). The shell of a hawksbill is incredibly beautiful. Serrated on the bottom with shades of orange, brown, yellow, and red, before becoming protected they were prized for jewelry and trinkets, which has placed their populations at risk.

FIGURE 13-3:
A hawksbill.

Source: Cristina Mittermeier – www.sealegacy.org

Loggerheads

Loggerheads (see Figure 13-4) have big heads and powerful jaws to crush prey, such as sea urchins, mollusks, and horseshoe crabs. Soon after they hatch on the sandy Florida shores, they swim out to a patch of sargassum (a type of floating seaweed) in the Sargasso Sea, where they spend their early years. The sargassum serves as a camo blanket, hiding the babies and soaking up the sun's rays to keep them warm. The warmth increases the baby loggerhead's metabolism to help fuel its growth.

Flatbacks

Flatback sea turtles . . . well, you guessed it, have flat shells, which are also thin and easily damaged. They live only in Australia, and they prefer turbid water. They feed on squid, sea cucumbers, soft corals, and mollusks.

Kemp's ridleys

Kemp's ridley, named after Richard Kemp, who discovered the species and was first to study it closely, is the smallest of the sea turtles. Kemp's ridleys are found in the Gulf of Mexico and the east coast of the U.S. and Canada.

Source: Cristina Mittermeier – www.sealegacy.org

Both Kemp's ridleys and olive ridleys are the only two species of sea turtles that nest in large groups — a practice called *arribada nesting* (c'mon gals, let's hit the beach!). Hundreds, sometimes thousands of female ridleys storm the beaches all at the same time to lay their eggs (see Figure 13-5). Arribada nesting is thought to give the nesting mothers and the hatched babies a greater chance of surviving predators.

Source: Conor Goulding, © Mote Marine Laboratory

Like loggerheads, baby Kemps head to the Sargasso Sea to rest and grow. As soon as they're large enough, they develop a taste for crabs (their favorite food).

Olive ridleys

Olive ridleys are a little larger than Kemp's ridleys, and they get their name from their olive-colored, heart-shaped shells. Like Kemp's ridleys, mama olive ridleys engage in arribada nesting.

Making babies

Sea turtles become sexually mature at different ages depending on the species: 7–13 years for leatherbacks, 11–16 years for ridleys, 20–25 years for hawksbills, 25–35 years for loggerheads, and 26–40 years for green sea turtles. They mate in the water, and several weeks later, the female crawls onto a beach, digs a hole, lays her fertilized eggs, and covers them up. One female can lay about 200 eggs, which hatch about two months later. As soon as baby sea turtles emerge from their nest they race to the relative safety of the water to begin their lives at sea. Honestly, they're about the cutest babies to hatch from eggs, *ever* (see Figure 13-6), except for maybe a duckbill platypus, which may tie for that honor.

apiguide/Shutterstock

FIGURE 13-6: A baby green sea turtle.

Unfortunately, instead of being coddled, many of these babies become a tasty treat for birds, fish, crabs, and even mammals (but that's nature). Consequently, only an estimated 10 percent of baby turtles normally reach adulthood, and if threats from humans are added in, that number may be as low as 1 percent.

A seriously threatened animal

REMEMBER

Except for the Australian flatback, all sea turtles are endangered. The modern world has given them too much to contend with — fishing nets; ghost nets (discarded fishing nets and other plastic debris they get tangled in); boats (see Figure 13-7); poachers (of turtles and their eggs), coastal development encroaching on their nesting sites; pollution, including light pollution; rising seas due to climate change; and, saddest of all, balloons, plastic grocery bags, and other plastic and rubber products that look like jellyfish and other prey but are indigestible (so the turtles die on a full stomach). Regardless of where you live in the world, sea turtles near you could use some help.

FIGURE 13-7: Turtle with boat scars; fortunately, this one survived.

Source: Cristina Mittermeier – www.sealegacy.org

One big threat to sea turtles is climate change. Rising sea levels threaten their nesting sites, and rising temperatures are on track to severely reduce populations of male sea turtles. The temperature of the nest determines the sex of the babies — females when the nest is warm, males when it's cool. Historically that wasn't a problem because varying temperatures ensured a balanced number of each sex were born, but as Earth warms, the concern is that not enough males will be born, dooming sea turtles to extinction.

Another problem is coastal development, which threatens sea turtle survival in two ways. First, it destroys their nesting sites, which are already at risk due to rising sea levels. Sea turtles are *natal nesters,* which means they return to the same beach where they were born to lay their own eggs year after year. When buildings are erected on or near beaches where these turtles have traditionally nested, they have nowhere to go. Early European settlers arriving in the Caribbean described beaches in the Bahamas and Cayman Islands as being so thick with nesting turtles that no sand could be seen. Now, virtually no turtles nest on these islands.

Coastal development also results in light pollution, which confuses the hatchlings. Ideally, baby sea turtles hatch at night, in the dark, which helps to protect them from predators as they scamper toward shore. How do they know where to go? Well, their main guide is light, so they head toward the brightest light source, which *should* be the moon reflecting off the ocean. Unfortunately, in developed areas, the brightest lights may be streetlights, lights in homes, or security lights. So instead of crawling into the ocean, baby turtles head toward these artificial lights and end up dying.

REMEMBER

To avoid confusing baby sea turtles during periods when they're due to hatch, turn off the outside lights and close your curtains. You can also use red lightbulbs for outdoor lighting, which allows you to see without confusing the hatchlings since they don't see red light.

Will the Real Sea Serpent Please Slither Forward?

Sea snakes are a group of ocean-dwelling snakes that number about 55 species of true sea snakes (subfamily *Hydrophiinae*) and six species of *kraits* (subfamily *Laticaudinae*). True sea snakes are venomous serpents (in the same family as cobras) specially adapted to live in marine environments. They have a vertically flattened body and a paddle-shaped tail that enable them to "slither" through water. Unlike land snakes, whose scales overlap, most seas snakes have tiny scales that don't overlap, and some have lost the scales on their bellies altogether. Sea snakes also have "scale organs," called *sensilla*, which appear to have mechanosensory functions (like the lateral line some fish have).

Other aquatic specialties? They don't stick out their tongues as far as other snakes do; just the tip is exposed to "smell" the water. In addition, because they're air breathers, they have elongated lungs that extend down their entire body for increased lung capacity, and their nostrils are positioned on the top of their nose (instead of at the front) to breathe more easily when they surface for air, which isn't as often as you might think for one particularly cool reason — sea snakes can breathe through their skin! They obtain about 30 percent of their oxygen through cutaneous respiration. The blue-banded sea snake can actually transport oxygen directly from the water to its brain. Talk about efficiency! Thanks to these specialized adaptations, sea snakes can spend an hour or more underwater, especially if they don't move much.

Sea snakes also have salt glands (under their tongues) but research suggests that these glands aren't terribly efficient and that sea snakes partially quench their thirst by drinking fresh water that pools on the surface of the ocean when it rains.

REMEMBER

When you're in the ocean, you don't need to waste time debating over whether a certain snake is venomous — *all* sea snakes are venomous. In fact, they're some of the most venomous snakes in the world. The good news is that they're usually very docile. Many fishermen handle them with bare hands when they get caught in nets and return them to the sea without getting bitten. However, exceptions always exist, so we recommend not handling them unless you know what you're doing. (They are in the cobra family, after all.)

Depending on the species, sea snakes eat small fish, eel, fish eggs, or small octopi, and they mate and give birth to live young in the water.

Six species of sea snakes, called *sea kraits,* are not considered true sea snakes because they're ill-equipped for ocean living. They have round bodies, belly scales, and lateral nostrils, and they lay eggs on land instead of giving live birth in the ocean (the big phonies!). See Figure 13-8.

FIGURE 13-8:
A banded sea krait pretending to be a sea snake.

Source: Keith Ellenbogen – www.keithellenbogen.com

The Only Lizard to Make the Cut: Marine Iguanas

Although turtles and snakes have made a pretty successful transition from land to sea, lizards haven't done so well. In fact, only one species of lizard calls the ocean home — the marine iguana (see Figure 13-9). This sole species inhabits the Galapagos Islands, which form an archipelago off the coast of Ecuador, South America. While these mini-Godzillas may look like menacing carnivores, they're actually vegetarians, feeding primarily on algae they gnaw off the rocks. Although iguanas prefer the rocky coasts, some like to hang out in mangrove swamps and on beaches as well.

FIGURE 13-9: A marine iguana sunning on rocks.

Source: Anamaria Chediak – www.anamariachediak.com

Being herbivores, these lizards eat algae, seaweed, and seagrasses along the shores on the islands. They can dive to depths of 20 meters (65 feet) and stay under for up to 30 minutes, but most of their undersea meals are shorter than that, and they spend a large part of their lives on land sunbathing to stay warm.

Marine iguanas have had to evolve over time to become suited to this seaside existence. Specifically, they needed long, flat tails for swimming, long claws for hanging on to rocks, stubby snouts for grazing on seaweed and seagrass, and some way to get rid of all that salt in their diets. If you were to compare a marine iguana to a land-based species, the first three adaptations would be readily evident. The marine iguana's tail is much flatter and more powerful, its claws are longer and stronger, and its snout is stubbier.

The fourth adaptation (the ability to secrete salt) is just as obvious, assuming you know what you're looking for. When you examine most marine iguanas, you notice that the very tops of their heads are white. This white cap isn't a sign of old age; it's actually a crust of salt. Like other marine reptiles, the iguana has a special gland that removes excess salt from its system. This gland is located between the iguana's eyes and nostrils. When enough salt accumulates, the iguana blows the buildup out its nostrils, and the discharge typically settles on its head. Marine iguanas often blow out a spray of the snotty salt to ward off any annoying intruders too (a big salty sneeze will make just about anything back off!).

Saltwater Crocodiles

Saltwater crocodiles (often called "estuarine crocodiles" or "salties") are the largest living reptiles. Shown in Figure 13-10, they have a long body, short legs with webbed feet, a long and toothy snout, a powerful tail, armored skin (covered by scutes), powerful jaws, and salt-secreting glands (in their mouths) that make them as much at home in saltwater as in freshwater environments. Most salties range in length from 4 to 5 meters (13 to 16.5 feet), but some people have reported crocs double that size. An adult male can weigh in excess of 998 kilograms (2,200 pounds). They frequent estuaries, lagoons, and mangroves, but some also spend a considerable amount of time at sea riding the currents or in rivers or other freshwater environments. If you're interested in meeting one face-to-face (though we don't recommend it), you can find them hanging out in and around northern Australia, Papua New Guinea, and parts of Southeast Asia. Only one other species of crocodile, the American crocodile, also tends to prefer saltwater. They live in Florida, the Caribbean, and Central and South America. They are not quite as large or as ill-tempered as their Australasian cousins.

Crocodiles are ferocious predators with powerful jaws capable of clamping down at pressures of 5,000 pounds per square inch, in contrast to 100 pounds per square inch for the human jaw. However, they have hardly any opening mouth strength — their mouths can be held shut with a strong rubber band — so chewing can be a challenge! To adapt, they've developed a couple techniques to aid their digestion. The first is a death roll that's used to both kill their prey and dismember them to make them easier to swallow and digest (oh how nice). The second is that they may swallow small rocks to help crush and grind up their food inside their stomachs.

FIGURE 13-10:
A saltwater
crocodile.

They have a very slow metabolism, so they can go months between meals. When they do eat, their favorite food seems to be . . . well, just about anything that gets close to the waterline — mammals, birds, fish, and even other crocodiles. Trust us, saltwater crocodiles are nothing to mess with and have earned their reputation as cranky critters. However, like all creatures, they play a critical role in the environment, so even though they can be scary, they're essential members of the community.

Despite their fearsome reputation, courtship between crocodiles is surprisingly gentle. After a female and male have decided they're suitable for one another, they engage in a complex ritual of nuzzling their heads and bodies together before mating in the water (aww). This courtship can last for several days before they mate.

If that isn't enough to soften their reputation, consider that female crocodiles are actually quite motherly. They lay somewhere around 50 eggs in a nest of decomposing leaf litter near the water's edge during the rainy season. About 12 weeks after the eggs hatch, the mother helps them to the water (sometimes carrying them gently in her mouth) and then watches over them for a few more weeks until they're big enough to head out on their own. Mama crocs even communicate with their babies using gentle chirps and come running to protect their offspring the moment they hear their babies cry for help (so don't mess with the babies, either).

FUN FACT

It's true, crocodiles cry genuine crocodile tears. They secrete water over their eyes to keep them moist and clean. Even cooler is that these tears are a favorite beverage for butterflies.

ALLIGATOR OR CROCODILE?

To the untrained eye, alligators and crocodiles look nearly identical. They both have a long body covered with hard scales, short legs, a powerful vertically flattened tail, and a long snout with lots of teeth.

However, alligators are usually grey and black, whereas crocs are olive or tan. An alligator's snout is shorter and wider than that of a crocodile, and their "smiles" are different. With alligators, you're not likely to see their pearly whites unless they open their mouths, whereas a croc's teeth are usually visible even when its mouth is closed.

They also differ in where they like to hang out. Alligators live exclusively in freshwater lakes, rivers, and swamps, though they can tolerate salt water for a few hours at a time. Crocodiles live in both freshwater and saltwater environments though they usually prefer salt water. While alligators often sunbathe on shore, crocs spend far more time in the water.

And though crocs are more aggressive than alligators, we wouldn't recommend picking a fight with either one.

FUN FACT

As with sea turtles, the sex of a crocodile is mostly determined by the temperature of the nest in which the eggs were laid. For instance, for Nile crocodiles, temperatures below about 31.7 degrees Celsius (89 degrees Fahrenheit) produce females, temperatures between 31.7 and 34.5 Celsius (89 and 94 Fahrenheit) produce males, and above 34.5 Celsius (94 Fahrenheit) produce females.

IN THIS CHAPTER

» **Determining whether a bird qualifies as a seabird or a shorebird**

» **Plunging the depths with penguins, loons, grebes, and other diving fowl**

» **Getting the scoop on pelicans and other Pelecaniformes**

» **Hugging the shoreline with ospreys, heron, and flamingos**

» **Bumping into sea ducks, gulls, geese, and other fascinating ocean fowl**

Chapter **14**

Bird Watching in and Near the Ocean

M arine ecosystems extend above the surface of the ocean and beyond the coastlines to encompass creatures that spend much of their lives out of water, namely seabirds and shorebirds/waders. These waterfowl, adapted to thrive in marine environments, include a broad assortment of curious creatures including penguins, which use their wings to swim instead of fly; pelicans, which drop from the sky to scoop up prey in their pouchy beaks; the majestic albatross, which can fly thousands of miles over open seas; puffins — clownish looking birds that nest on the rocky cliffs and spend long periods at sea; flamingos — strikingly pink and highly skilled at standing on one leg while eating upside down; and hundreds of other species of birds that have adapted to seaside living.

This chapter explores the wonderful world of marine waterfowl and introduces you to an assortment of their diverse population.

Knowing What Makes a Bird a Shorebird or a Seabird

Before we get into any discussion of the difference between seabirds and shore-birds, consider what makes a bird a bird. To be a bird, you must:

>> Be a *vertebrate* (an animal with a backbone)

>> Be *warm-blooded* (able to control your body temperature)

>> Have feathers

>> Have wings, though not all birds fly

>> Have a beak

>> Lay eggs

>> Have a respiratory system through which air flows in a continuous one-way path through the lungs and back out again (an awesome adaptation everyone should have!)

FUN FACT

A bird's lungs don't expand and contract like ours to inhale and exhale. Instead, it has nine air sacs that inflate and deflate to push air through the lungs along a continuous path. With this system, oxygen flows into the lungs while carbon dioxide flows out continuously, whether the bird is inhaling or exhaling.

In addition to being birds, seabirds and shorebirds share the following traits:

>> **Depend on the ocean:** They live, feed, and breed in or near the ocean and depend on it for most of their food.

>> **Secrete salt:** They ingest salt when they eat and drink and most have salt-secreting glands, typically located in the bird's beak or near its eyes.

>> **Nest, breed, and lay eggs on land (or in trees):** Even if they spend most of their lives at sea, they must return to land to breed, lay eggs, and nest.

Marine birds can be divided into two groups — those that hang out mostly on shore and those that spend most of their time at sea.

Shorebirds

Shorebirds hang out along the coastlines — on beaches or rocky shores, on rocky cliffs, or in intertidal flats, estuaries, mangroves, or wetlands (see Chapter 5 for more about these marine ecosystems). They tend to feed in shallow water, usually on marine worms, mollusks, crustaceans, and small fish and, because of this, are often called *waders* (although some classification systems distinguish shorebirds from waders). Many are migratory species, travelling long distances between seasonal habitats to feed or breed, while others prefer to stay put. While some species form large flocks, most are solitary nesters.

Shorebirds are a group of about 400 species that include sandpipers, plovers, avocets, herons, egrets, flamingos, oystercatchers, ibis, rails, and spoonbills.

Seabirds

Seabirds hang out on the shorelines with shorebirds, but they primarily head out to sea to feed on fish, squid, crustaceans, and other small animals that live in the surface waters of the open ocean, returning to shore to nest or take care of their young. Many seabirds, such as the albatross, are long-distance travelers and can spend many days flying thousands of miles and feeding along the way before returning to land. More than 96 percent of seabirds are colonial, which means they form large groups to breed.

Seabirds account for about 350 species, including albatross, frigates, fulmars, petrels, shearwaters, and tropicbirds. Puffins, murres, boobies, auks, and penguins are also included in this group.

PRESSURED POPULATIONS

Unfortunately, many seabirds and shorebirds are considered threatened or endangered. By some estimates, populations of migratory shorebirds have dropped by 70 percent in North America since the 1970s, while seabird populations have declined by 47 percent — 31 percent are listed as threatened, endangered, or critically endangered, making them the most threatened group of birds.

Invasive species, bycatch, overfishing, disturbance while nesting, coastal development, and climate change are all reasons these birds are declining in numbers. Additionally, as many seabirds and shorebirds cross international boundaries while migrating or foraging, protecting them is a complicated challenge. Arctic terns, for example, fly from the Arctic to the Antarctic every year and (you thought you had a lot of frequent flier miles), so widespread international cooperation is needed to protect these birds across their entire migration range.

International agreements and legislation, such as the Chinese-Australia Migratory Bird Agreement (CAMBA) and the Migratory Bird Treaty Act, are helping to protect birds as they migrate and preserve their rest-stop habitats. Another crucial agreement is the Ramsar Convention on Wetlands, which aims to conserve the wetland habitats that many migratory species rely on.

Flying Way Below the Radar: Penguins

No seabirds are more committed to ocean living than penguins. Their wings have even evolved to be more like flippers. They can't fly, and they can barely walk, hopping or waddling across the ice like awkward waiters in tuxedos or sliding around on their bellies. But when they hit the water, they're some of the most graceful seabirds on earth, using their strong wings to propel themselves while steering with their tails and webbed feet. Some can swim up to 35 kilometers (22 miles) per hour in short bursts, plenty fast enough to catch their prey — mostly krill, small fish, squid, and cuttlefish. They're also respectable divers. Some emperor penguins (see Figure 14-1) have been recorded catching prey at depths of 565 meters (1,854 feet), though penguins rarely dive much deeper than about 200 meters (650 feet).

FIGURE 14-1: Emperor penguins with chicks.

Source: Paul Nicklen – www.sealegacy.org

Penguins are highly adapted to an aquatic lifestyle in frigid climates. Like dolphins and most fish, the penguin's body is tapered, reducing resistance underwater. Unlike most birds, which have hollow bones, the penguin's bones are solid, reducing its buoyancy, so it can dive more easily. Its wing bones are fused and covered with tough skin, making them more like paddles. Their feet are positioned toward the rear of their bodies, near the tail, serving as a rudder to steer underwater. The penguin has a double-insulated hide, consisting of a thick layer of fat on the inside covered by skin and a densely packed layer of feathers shaped more like needles than true, branching feathers. Pockets of air can be trapped within this layer of feathers, allowing the penguin to remain buoyant, while oils rubbed into the feathers help them stay waterproof.

Why wear a tuxedo in the water? The tuxedo is an example of *countershading* — a cloaking mechanism that makes them harder to see in the water. A white belly, when viewed from below, disappears against the bright background of the sky, while a dark back, when viewed from above, disappears against the darkness of the ocean depths. Many seabirds use countershading as their personal camo.

Roughly 17 species of penguins (up to 20 by some counts) live primarily in the Southern Hemisphere, from Antarctica to the coasts of South Africa, South America, New Zealand, and Australia. These include the Adélie and emperor penguins, which stay in Antarctica year-round, as well as the king and rockhopper (among others), which live in the subantarctic islands, and the gentoo, macaroni, and chinstrap penguins, which are found in both areas at different times of year.

The only exception to this rule is the Galápagos penguin (see Figure 14-2), whose range extends slightly across the equator into the Northern Hemisphere. (Water near the Galápagos Islands is cooled by the Humboldt Current, making it tolerable for the cold-loving penguins.) Some penguins travel long distances to reach their feeding grounds, and for some species of penguins, foraging trips can last weeks.

FIGURE 14-2:
Galápagos
penguin.

Source: Anamaria Chediak – www.anamariachediak.com

FEELING THE HEAT

Because penguins thrive in icy cold water, they're increasingly threatened by global warming, which reduces their habitat, disrupts food webs, causes large fluctuations in ice levels, and even increases rain, washing away nests and eggs and causing chicks to freeze and die. Warmer conditions are especially difficult for southern penguins such as emperor penguins, which rely on the ice to breed and rear their chicks. That icy cold water also provides an essential habitat for their food — the krill that they (and their other prey) feed on.

However, the problem goes beyond melting ice; fluctuating ice levels can be just as harmful — when particularly warm years cause the sea ice to become more spread out, but not as thick, penguins have to travel farther to find food. At least one study predicts an 86 percent decline in the populations of Emperor penguins by 2100 as a result of climate change.

There is hope, though. In the last few years, hidden colonies of penguins have been found, including 11 new colonies of Emperor penguins in 2020, boosting their known population by 20 percent! These colonies were spotted from space, where satellite imagery picked up large patches of *guano* (brightly colored penguin poop) that ultimately gave away their location. Additionally, a previously known colony of Adélie penguins was found in 2018 containing roughly 1.5 million penguins (much, much more than initially thought), which increased the known population in the region by almost 70 percent. But, unless action is taken to stop climate change and mitigate its impact, these colonies will suffer as well.

Although they're skilled predators, penguins are prey for leopard seals, sharks, orca, and other larger animals, while some predatory birds, such as skuas and sheathbills, snatch penguin eggs and chicks. That's kinda messed up, but everyone's gotta eat.

Going Loony

Well-known to any Canadian with a $1 coin, loons (also called *divers*) are the Northern Hemisphere's answer to penguins. They may not look much like penguins, but they're sure built like them, with solid bones and their legs positioned way back on their bodies. Simply by exhaling and releasing the air trapped in their wings, loons can lower their bodies nearly level with the surface of the water. They can dive down hundreds of feet and stay submerged for several minutes, zigzagging expertly through the water to nab prey and elude predators. They have sharp beaks and rearward-pointing projections on the roofs of their mouths and on their tongues that enable them to firmly grasp slippery fish.

Unlike penguins, which have lost the ability to fly, loons are excellent aviators, able to achieve speeds of up to 96 kilometers (60 miles) per hour. However, with their dense bodies, solid bones, and small wings, take-offs and landings can be a little awkward. They have to drag their bodies through the water for several yards before they get up enough speed to lift off. They're also awkward on land, which is what earned them the name "loon" — because of their less-than-graceful way of walking.

FUN FACT

Mother loons lay only one or two eggs, and their babies often ride on their backs.

Courting Grebes

Grebes are a family of diving birds that resemble loons— they have sharp beaks, and their legs sit far back on their bodies. They also have similar prey, feeding on small fish and aquatic invertebrates. They have a dense layer of feathers uniquely structured — sticking almost straight up and curling at the ends. By ruffling or smoothing their feathers, they can adjust their buoyancy. Unlike loons, which have webbed feet, a grebe's feet are lobed — their toes are shaped like paddles without webbing connecting them. Though most grebes can fly, some species, such as the Titicaca grebe from South America, have lost that ability.

FUN FACT

Grebes are known for eating feathers (typically their own). Scientists are unsure why but think it has something to do with preventing internal injury from fish bones or ensuring their food is completely digested. But don't try this at home folks!

Grebes seem to prefer still or slow-moving bodies of fresh water (marshes, lakes, and ponds), but some species spend their winters in sheltered coastal areas (bays and estuaries). They build floating nests out of vegetation, and when the young hatch they're able to swim almost immediately.

Like many other birds, grebes have distinct breeding and non-breeding plumage and interesting courtship rituals. Prior to mating, a couple dances together, shaking, bobbing, or flicking their heads; diving; moving their wings; or displaying "penguin posturing," standing almost upright in the water. Some, such as the great crested grebe, dance with and present aquatic vegetation to each other. Kind of like giving your sweetheart a bouquet of flowers.

Tubular, Dude! Albatross, Petrels, Shearwaters, and Fulmars

Albatross, petrels, shearwaters, and fulmars represent about 117 species that have one big thing in common — tubular nostrils that project from their upper bill. Also, their feet are webbed, and their hind (fourth) toe is either missing or vestigial (nearly gone). One more thing — they all have a powerful musky odor from the discharge of stomach oil, something they can do on demand to ward off predators (lovely). In this section, we focus on the albatross and give a shout-out to the other members of the group.

Soaring with the albatross

Albatross are a group of a little over 20 species that patrol the skies over the Southern Ocean and northern parts of the Pacific (see Figure 14-3). They include the bird with the largest wingspan — the wandering albatross, whose wings span up to 3.7 meters (more than 12 feet). Albatross are also equipped with a locking system within the joints of their wings, which keeps them in the open position without requiring any expenditure of energy. Together with their huge wings, this locking system makes them exceptional gliders. They can remain airborne for months at a time, covering a distance of more than 15,000 kilometers (9,300 miles) in a single trip! That's roughly the equivalent of flying from New York City to Sydney, Australia, nonstop. They use two techniques to soar through the air without flapping their wings:

>> **Dynamic soaring:** They descend downwind and then turn into the wind, using their downward momentum and the updraft to generate lift, propelling them upward.

>> **Slope soaring:** They glide over the waves, using small updrafts produced by the waves to keep them aloft.

Albatross spend nearly their entire lives at sea, feeding primarily on fish and squid, but they can also scavenge for food or dine on crustaceans, zooplankton, and other small animals. Many albatross species are like sea turtles in that they return to land only to breed. They typically take about five to seven years to reach sexual maturity and don't start to breed until they're about ten years old, at which time, they choose a lifelong mate. How sweet.

Because they can live for more than 60 years, they're highly selective when choosing a mate. Courtship involves preening each other, dancing, and singing (ahhh, to be young and in love!). Every few years they return to the same colony where they were born to mate, a habit called *natal nesting.* The female lays one egg, then the partners take turns tending the egg/chick and hunting for food. The proud parents care for the chick for about four to nine months before it's ready to set out on its own. The young bird then spends the first few years of its life at sea, without touching land.

Source: Anamaria Chediak – www.anamariachediak.com

FIGURE 14-3:
An albatross.

FUN FACT

Albatross are generally a long-lived species, with the oldest known albatross being Wisdom, the Laysan albatross who's at least 66 years old and is still having chicks!

Skimming the surface with shearwaters

Other tube-nosers are the shearwaters, which have earned their name from flying so close to the water that they skim the tops of waves. In addition to being able to cover long distances in flight, like their larger cousins (the albatross), shearwaters are excellent divers, plunging to depths of up to 70 meters (230 feet) to nab their prey, typically fish and squid.

Shearwaters engage in huge, long annual migrations; for example, the short-tailed shearwater flies annually from southern Australia to Alaska, and the sooty shearwater flies an even longer distance from New Zealand or the Falkland Islands to Japan and Alaska and sometimes to the North Atlantic.

ALBATROSS UNDER SIEGE

Unfortunately, many species of albatross are listed as either endangered or critically endangered. Historically, they were hunted for their meat and feathers, but today, they face numerous threats, including:

- Rising sea levels, due to climate change, eroding their nesting grounds.

- Loss of prey species due to overfishing.

- The introduction of predatory species: Albatross traditionally have nested in areas void of predators, so they're more trusting than other birds. The introduction of rats (which eat eggs and chicks) and cats (which eat chicks and adults) to some colonies have led to declining albatross populations.

- Plastic pollution, which causes high mortality especially in chicks. Parents pick up the bright plastic fragments thinking they're food (they even smell like food due to the microorganisms they attract) and feed them to their chicks. The plastic clogs and damages their digestive systems or they feel full, stop eating, and starve to death with a full belly.

- Longline fishing: This commercial fishing technique involves one long fishing line with shorter lines attached to it at intervals, each with a hook on the end. Sadly, seabirds try to feed on the bait or on the fish caught on the hooks, get tangled in the line, and drown. Some longlines can be 130 kilometers (80 miles) long with up to 40,000 hooks. By some estimates, longline fishing kills up to 300,000 birds every year. As long-lived species, even a handful of albatross deaths per year can influence a population.

Some steps have been taken to limit the number of birds getting caught or tangled in longlines, such as introducing bird-safe hooks, fishing at night when seabirds aren't as active and feeding, using lines covered in streamers to scare birds away, and weighing hooks so they're deep enough in the water column that birds can't reach. However, further action is needed.

Fluttering above the surface with petrels

Petrels are generally smaller than either the albatross or shearwaters, and they often flutter just above the water's surface dipping down to feed on planktonic invertebrates. Many are referred to as "storm petrels" or "water witches" because their arrival is believed to predict approaching storms. The Wilson's storm petrel is in this group, which is considered by some to be the most abundant group of seabirds. The dovelike snow petrel (see Figure 14-4) lives in Antarctica and has one of the most southerly breeding sites of any bird.

FIGURE 14-4:
Snow petrel
in flight.

Source: Cristina Mittermeier – www.sealegacy.org

Pelicans and Other Pelecaniformes

Pelecaniformes (say that six times fast) are an order of six families comprising about 67 species of seabirds that include pelicans, boobies, gannets, cormorants, shags, frigate birds, and tropicbirds. What do all these diverse birds have in common? Well, for starters, they're all medium-to-large aquatic birds with *totipalmate feet*, meaning they have webbing connecting all four toes. Second, they all have an expandable *gular sac* that joins the lower mandible of the beak to the bird's throat. In this section, we shine the spotlight on pelicans and mention a few other representatives of this group.

FUN FACT

Some Pelecaniformes species have a couple additional adaptations that enable them to dive into the water to catch their prey from great heights without being injured. Pelicans, boobies, and gannets, for example, have strong, lance-shaped beaks; sinuous necks; inflatable air sacs in their face, neck, and breast to cushion the impact (like air bags in a car); and either no external nostrils or sealable nostrils that prevent water from entering their lungs during dives (yes, they're mouth-breathers).

Pelicans

Pelicans are a group of eight species of large iconic seabirds (and Philippe's favorite), mostly white (but some brownish), well-known for their long beaks and spacious throat pouches (again, a *gular sac*) used for feeding (see Figure 14-5). If you've ever repeated the rhyme, "A wonderful bird is the Pelican. His beak can hold more than his belly can," you were scientifically accurate. The throat pouch of a pelican does hold more than its stomach — up to 13 liters (more than 3 gallons) in some species. The bill is also structured in a way that enables the pouch to expand sideways to accommodate larger prey.

At first glance, you might doubt that a pelican could fly. Its large body, thick neck, and massive beak don't exactly cry out "ready for takeoff," but they're actually quite graceful in the air, gliding in formation near the ocean's surface or riding the higher altitude winds for long distances. They have different seasonal home ranges, flying wherever necessary in pursuit of food.

Source: Claudio Contreras-Koob – www.wildcoast.org

FIGURE 14-5:
Two pelicans roosting in a tree.

They feed primarily on fish but will dine on other animals, including turtles and other birds, when the opportunity presents itself. They're not deep divers, but when they spot prey, they plunge beak-first into the ocean at speeds of up to 65 kilometers (40 miles) per hour, wings back to prevent damage, then open their beaks to scoop up their meal (see Figure 14-6). The tip of their beak is shaped like a hooked tooth, preventing prey from escaping. The pelican then siphons off the water and manipulates the food so it slides down its throat head-first. (One type of pelican, the brown pelican, prefers to dive-bomb its prey with mouth open.)

FIGURE 14-6:
Pelicans diving
for their prey.

Frigate birds

Frigate birds are a family of five species of seabirds that live in tropical and sub-tropical regions. They're mostly black with a long, forked tail; long, angular wings; hooked bill; and the characteristic red, inflatable gular sac that develops during the breeding season (see Figure 14-7). They lack oil glands to waterproof their feathers, so they engage more in skimming than in diving for their food — fish (especially flying fish), squid, jellyfish, and even baby turtles. They're also known to steal food from other seabirds, such as pelicans and boobies.

Boobies and gannets

Boobies are more than just funny-looking birds with a silly name (and Ashlan's favorite bird); they and their relatives, the gannets, are impressive divers with strong necks and bullet-shaped bodies (see Figure 14-8). They dive from great heights in pursuit of prey. Some boobies have been known to dive from a height of 100 meters (more than 300 feet), while some gannets reach a speed of 100 kilometers (60 miles) per hour before they hit the surface. They frequently hunt in flocks, dropping from the sky like a flurry of arrows when they spot a school of fish, each surfacing with a fish wedged between its scissorlike jaws. Blue-footed boobies and red-footed boobies are known for their fabulous colored feet — and their funny way of dancing with said feet when trying to show off for the opposite sex.

Source: Anamaria Chediak – www.anamariachediak.com

Source: Anamaria Chediak – www.anamariachediak.com

Gannets prefer a cool, northern climate and nest in rocky areas and cliffs. Boobies prefer more tropical regions. A booby's eyes make it look kind of dumb, but the name "booby" was probably derived from the bird's habit of landing on the decks of ships and allowing sailors to capture them easily (maybe they just wanted to say hi?). Both boobies and gannets are highly social birds (what did we tell ya!), especially when it comes time to breed and lay eggs. Both are known to gather in colonies numbering in the tens of thousands.

Cormorants and shags

Like gannets and boobies, cormorants dive for their food, but they don't dive-bomb. They're built more like loons and swim like them, too. They have dense bodies that sit low in the water and enable them to swim underwater for long distances in pursuit of their prey. You can find a great video online of a cormorant yanking a remora (suckerfish) off the side of a whale shark.

Unlike boobies and gannets, cormorants aren't exclusively marine, and they're not as waterproof, because their oil glands aren't as well-developed. As a result, they must spend more time out of water drying their wings before they can take flight again after feeding. They often perch on rocks with wings spread wide to dry (see Figure 14-9). The flightless cormorant has given up on flying altogether; its wings are so short it can't even get airborne.

FIGURE 14-9: A double-crested cormorant drying its wings.

Photo by Colin Durfee. Licensed under CC BY 2.0

Shags are nearly identical to cormorants, except they're smaller and have thinner beaks, less white and yellow where the beak meets the throat, and they tend to kick their legs up in the air when they start to dive, whereas the cormorant slips gently down below the surface.

Phaethontidae — Tropicbirds

Tropicbirds are three species of large seabirds, up to 1 meter (3 feet) in height, that live in the Atlantic, Pacific, and Indian Oceans. They're graceful flyers, predominantly white with long central tail features and spindly legs (see Figure 14-10). Once thought to belong to the Pelecaniformes order, they are now classified as their own order. But like Pelecaniformes, tropicbirds plunge-dive to capture their prey.

FIGURE 14-10:
A red-billed tropicbird.

Source: Anamaria Chediak – www.anamariachediak.com

Sea Ducks and Geese: The Saltwater Variety

When you think about birds that live in and around the ocean, the first bird that pops into your mind probably isn't a duck or a goose. Those are freshwater birds, right? Well, most are, but some prefer saltier environments either as their main habitat or as rest stops along their migration route. These include about 15 species of sea ducks, such as eiders, scoters, goldeneye, bufflehead, and mergansers. Many form large flocks along the coast, and they feed on crustaceans, mollusks, echinoderms, aquatic insects, fish, or vegetation. Most of these species live in the Northern Hemisphere, in Asia, North America, Europe, Greenland, and Russia, and many migrate to the Arctic during the summer.

Some geese also frequent salty environments, including the snow goose and, especially, the brent (or brant) goose, a small goose that lives in estuaries during the winter and feeds on sea lettuce. They used to eat eelgrass, but a disease caused the die-back of eelgrass, pushing brent geese to the brink of extinction. They survived by switching to sea lettuce, which until that time had been a very small part of their diet, and they've been eating it ever since. They breed on the high Arctic tundra and winter along the coasts in more temperate zones.

A Curious Mix: The Charadriiformes

Accounting for more than 350 species of seabirds and shorebirds, the *Charadriiformes* are the most diverse group covered in this chapter. This order includes 50 species of gulls; 44 species of their closest relatives, the terns; 22 species of auks; 8 species of skuas; 3 species of puffins (which look more like parrots than gulls); and more than 200 species of shorebirds and waders.

Looking at the members of this order side-by-side, you would never think they were related; however, they share many internal anatomical traits, such as the structure of the skull and spinal column, leg tendons, *plumage* (feathers), and *syrinx* (voice box). Most of the external features vary from species to species. The following sections reveal the diversity of this group by examining some of its more common, interesting, and straight-up cute members.

Gulls, terns, skimmers, and friends

Gulls, terns, skimmers, skuas, and jaegers account for nearly 100 species of seabirds with wide wingspans and webbed feet that live in coastal areas and feed mostly near the surface of the ocean. As a group, they span the globe from tropical to temperate to polar regions.

Gulls

Gulls, also known as seagulls, are some of the most familiar faces at the beach. They're opportunistic birds that eat just about anything, which may explain why you can find them just about anywhere — from polar to tropical regions, from saltwater to freshwater environments, and from coastal to inland areas.

Although they're strong flyers, they rarely venture out to sea for long. They paddle and feed in the intertidal zone or shallows, or wait for other predators to drive prey to the surface, which the gulls then scoop up. They are highly vocal and prefer to remain in large groups — some gulls will even display mobbing behavior if predators or threats come too close to their nests. Gulls usually return to the same breeding site each year, and some gull species pair for life.

Terns

Terns are gull-like seabirds with streamlined bodies, longer beaks, forked tails, and thinner wings. Terns generally feed on zooplankton, fish, and small invertebrates by hovering above the water first, then diving, and can often be found where other large marine predators are hunting. They have a worldwide distribution, with some species living in wetlands or other coastal areas while others preferring life on the open ocean.

FUN FACT

Terns form large breeding colonies and often migrate together. Before a tern colony leaves a site, they fall silent for a brief period of time. This quiet moment is called a "dread," then all at once, they take to the air in a dramatic display and fly off. Isn't nature wonderful?

The small Arctic tern holds the record for the longest bird migration (and one of the longest animal migrations on earth), moving between the high Arctic and the Antarctic, hopping from one feeding ground to the next each year. They do it to follow the sun — talk about an endless summer! It's possible that terns experience more sunlight than any other animals because in the summer each pole has 24-hour sunlight. It's thought they do this because it is easier to hunt in sunlight, and the seas are calmer in the summer for these small birds. They travel about 70,000 kilometers (more than 43,000 miles) between breeding seasons. Over the course of their average 30-year lives, they're thought to fly the equivalent of three trips to the moon and back! They also mate for life.

Many terns are declining in population due to several factors, including coastline development, pollution, disturbance of colonies, introduced species predation, and illegal egg collection. The critically endangered Chinese crested tern in particular is thought to be down to 50 individuals. Monitoring and conservation efforts have been showing promising results, with the species being found in new locations and existing breeding colonies successfully rearing chicks.

Skimmers

Also known as scissorbills, skimmers are bizarre ternlike birds, best known for their underbite — the bottom of their bill is longer than the top. They use this to detect prey while fishing along rivers or coastlines by extending the lower part of their beak into the water as they fly close to the water's surface. As soon as the beak strikes prey (such as a small fish or crustacean), it snaps shut. They have catlike eyes with slit-shaped instead of circular pupils, which may help them avoid being blinded by the sun's glare or reflection off the water or help them see better during their usual feeding times (night, dusk, and dawn).

Skuas (jaegers)

Skuas (also called jaegers, which is German for "hunters") are considered to be the falcons of the ocean. They're large predatory or omnivorous birds — eating chicks, eggs, fish, *carrion* (decaying flesh of dead animals), small mammals, insects, or other birds. Sometimes known as "avian pirates," they're well-known for robbing other birds of their prey, a practice called *kleptoparasitism*.

Some skuas can be found close to the North and South Poles. The Arctic skua in the north (also known as the "parasitic jaeger") lives mostly at sea, returning to the northern coastlines of North America and Eurasia to breed. The south polar skua breeds along the rocky, ice-free coasts of Antarctica, where it has earned a bad reputation for stealing and eating penguin chicks and eggs (not cool, but they have babies to feed too). They are tough birds with some individuals even sighted as far south as the geographic South Pole where it gets really, really cold. While associated with the Arctic and Antarctic respectively, both species migrate across the equator at different times of the year.

Auks, puffins, and other Alcids

Alcids are a family of about 25 species of compact birds with short wingspans that are more comfortable "flying" underwater than in the air. Similar to penguins, they use their wings like flippers to propel themselves in the water and their feet like rudders to steer. Unlike penguins, they can also fly, and some fly long distances to migrate or to disperse after breeding.

Auks

Auks are black and white with compact bodies, short wings, and short legs with webbed feet. They nest in rock cavities or in burrows along coastal cliffs and spend their non-breeding season offshore feeding on fish, squid, and krill.

The great auk was once widespread across the North Atlantic, where it bred on isolated, rocky islands, but it went extinct as a result of hunting for its meat, feathers, fat, oil, and eggs. Once supposedly numbering in the millions, the last breeding pair was killed in 1844, and the egg they were incubating was smashed under a boot. *Ugh!*

Guillemots and murres

Guillemots and murres are true seabirds, returning to land only to breed. They look and move like penguins, jumping and waddling on land in their tuxedo-colored plumage, and using their short, powerful wings to dive, easily reaching depths of nearly 200 meters (650 feet). They nest on islands in the Atlantic Ocean and on the rocky coasts of England, Iceland, and Norway. During the winter, they swim to the southern North Sea to feed.

Puffins

Puffins are the most recognizable of the alcids, probably because they are so freaking *adorable,* especially when their bills turn orange during the breeding season (see Figure 14-11). Three species of puffin — the Atlantic, horned, and tufted — all live in the Northern Hemisphere.

FIGURE 14-11:
A tufted puffin.

Source: Romona Robbins – www.RomonaRobbins.com

In the past, puffins had disappeared from certain islands due to hunting for their eggs, feathers, and meat, but this prompted puffin lovers to launch reintroduction programs in those areas. A key example is the colony on Eastern Egg Rock in the United States, where *pufflings* (baby puffins, how cute is that?) were reared on the island to encourage them to return as adults. The strategies employed in these programs, such as using decoy birds to encourage other birds to land and nest, have been used and expanded upon to protect another 49 seabird species in 14 countries.

Tragically, the iconic puffins of Iceland are not faring so well. The Bering Sea is experiencing unprecedented warming and is causing the entire ecosystem to change and the fish puffins prey on to migrate north to cooler waters. Sadly, these waters are often too far for the puffins to fly, and they're either starving to death or not making it back to the islands in time and perishing during flight.

Shorebirds and waders

Shorebirds and waders account for about 220 species of birds that live primarily close to shore in coastal areas, walking or wading to feed. They include sandpipers, plovers, lapwings, snipes, stilts, godwits, curlews, jacana, and sheathbills. We cover several of these groups in the following sections.

Sandpipers

Sandpipers are small-to-medium-sized birds that tend to have long bills, long legs, narrow wings, and short tails that spend most of their time on beaches, mudflats, and inland bodies of water. They feed along the water's edge on insects, small crustaceans, and worms and are often recognized by their thin, piping cries. Many of them nest in the Arctic or in subarctic regions, building their nests right on the ground, and then winter in temperate regions from north of the equator to far south of the equator.

The sandpiper family also includes *godwits* — larger wading birds with long straight bills (see Figure 14-12) — and *snipes,* which have short legs and long, straight, flexible bills perfect for probing the sand for worms.

Stilts and avocets

Stilts and avocets have long legs and bills (the bill of a stilt is straight, while the bill of an avocet is curved slightly upwards), delicate necks and faces, and slender bodies. They feed in shallow waters in the intertidal zone, wetlands, and salt lakes. To find their prey, they dip their long bill underwater and move it side to side, or probe the sediment for crustaceans, insects, mollusks, and other invertebrates.

Most species have brightly colored legs and black-and-white plumage, with some avocets also having a rust-colored head.

Source: Claudio Contreras-Koob – www.wildcoast.org

FIGURE 14-12:
Godwits.

Plovers

Plovers are smaller shorebirds, with shorter legs and bills and plumper bodies. They usually feed on animals found under the sediment along sandy beaches or tidal mudflats, such as mollusks, worms, and crustaceans. Unlike other shore-birds, many hunt by sight rather than by touch. They are also known for their silly way of walking — sprinting forwards with their legs moving back and forth really quickly, then abruptly stopping, then starting again.

Oystercatchers

Oystercatchers are shorter, stocky (usually black or black-and-white) birds with thick legs and a long red-orange bill (see Figure 14-13). Depending on the species, they can be found both inland, where they feed on worms and insects, or along the coast. Their bills are strong to allow them to pry open bivalves such as mussels, clams, and oysters (hence their name). Oystercatchers feeding along coastlines also eat gastropods, polychaete (pronounced poly-keet) worms, and occasion-ally fish and crustaceans. Some oystercatchers also display *brood parasitism* — like cuckoo birds, they lay their eggs in other birds' nests, to be raised by birds which aren't their parents, in this case, the nests of gulls. Now *that's* some lazy parenting.

FIGURE 14-13:
An
oystercatcher.

Source: Anamaria Chediak – www.anamariachediak.com

Sheathbills

Sheathbills are the only bird family endemic to Antarctica and the subantarctic. They resemble a cross between a pigeon, chicken, and seagull, but their coastal lifestyle classifies them as shorebirds. They're also the only Antarctic bird not to have webbed feet. Despite their clean appearance, sheathbills are junk food junkies, eating just about anything, including penguin eggs, the food regurgitated by penguins for their chicks, carrion, blood, seal afterbirth, and penguin poo (yum, yum). Like some other types of shorebirds, they have a sharp spur on their feet which they use to defend their territories during the breeding season.

Ospreys, Herons, Flamingos, and Other Seaside Attractions

Many other bird species rely on marine environments but aren't included in any of the groups already covered in this chapter. Some of these species aren't technically classified as seabirds, shorebirds, or waders, but they certainly deserve a mention in the context of any discussion of birds that live in, on, or around the ocean.

Osprey

Osprey are large birds of prey, commonly confused with eagles or falcons, that frequent both marine and freshwater habitats (see Figure 14-14). After the peregrine falcon, ospreys are thought to be one of the most widely distributed land birds. They feed almost entirely on fish, which comprises about 99 percent of their diet. They have large, sharp talons and specialized pads on their feet for catching and holding slippery prey, and a reversible toe that enables them to grip fish more firmly.

FIGURE 14-14:
Two nesting
ospreys.

They have a wingspan up to 1.8 meters (6 feet), and their keen eyesight enables them to pinpoint the location of fish from far away. They usually snatch fish swimming near the surface, but occasionally dive below the surface in pursuit of their prey. They have valves in their noses to prevent water from getting in, and countershading which makes it harder for the fish to see them soaring above.

FUN FACT

Male osprey engage in a courtship display called a "sky dance" or "fish dance," holding a fish or nesting material in their talons while repeatedly diving and hovering while calling out to their prospective mate. Each pair constructs a huge nest that they typically use for several years.

Eagles

Other impressive birds of prey that utilize marine ecosystems are the sea eagles, which includes the fish eagles and the famous bald eagle. They're skilled hunters and (depending on the species) feed on fish and other aquatic animals, birds, reptiles, and small mammals. They also feed opportunistically on carrion and will steal food from other birds, such as osprey.

Sea eagles are generally huge, the largest of which is the Steller's sea eagle from northeast Asia, with a wingspan of up to 2.5 meters (8 feet). The bald eagle, white-bellied sea eagle, and white-tailed eagle are the largest in North America, Europe, and Australia, respectively. These eagles can be found either along coastlines or inland near wetlands, lakes, and rivers. While most species aren't considered threatened, the Madagascan fish eagle is listed as critically endangered, with fewer than 120 breeding pairs. They're thought to be declining as a result of competition with humans for fish, entanglement in fishing gear, and destruction of nests and habitat.

Herons

While not considered marine, herons (including egrets and bitterns) are another familiar face along the coast, wading in wetlands, estuaries, and shallow marine waters. They're well-known for their long legs, necks, and beaks and their ability to remain perfectly still in the water before quickly striking to capture fish or other prey (see Figure 14-15). They also hunt by shuffling their feet to disturb anything resting on the bottom and then nabbing them. The black heron, engages in "canopy feeding," which involves spreading its wings to shade the water making it easier to spot fish and other prey. Most herons feed during the day, but the night herons, as their name suggests, prefer to hunt in the dark or at dawn and dusk.

Source: Anamaria Chediak – www.anamariachediak.com

FIGURE 14-15:
A heron.

They also have something called *powder down,* which are feathers that produce a feather dust when they break down, which is believed to help with cleaning, waterproofing, and preventing parasites.

Flamingos

While more commonly found in alkaline or salt lakes and mudflats, flamingos are also colorful, showy, occasional visitors to coastal lagoons, mangroves, and wetlands (see Figure 14-16). Depending on the species, they feed on diatoms and other algae, small crustaceans and mollusks, and small fish. Their famous pink color comes from the carotenoid pigments in their food, which are then absorbed by fats in the liver and deposited in their feathers and skin. That's right, flamingos start life as fluffy grey chicks and would naturally be white but turn their signature pinkish-orange color because of their food. It's thought that being brightly colored may make certain birds appear more attractive as mates.

FIGURE 14-16:
Flamingo.

Source: Anamaria Chediak – www.anamariachediak.com

Flamingos are *filter feeders* — they stir up the sediment with their webbed feet, dunk their beaks into the slurry upside down, suck it in, then squeeze the water out the sides of their beaks and swallow the food that remains. They move locations (somewhat erratically) in response to changing environmental conditions and food availability.

IN THIS CHAPTER

» Checking out the unique traits of marine mammals

» Discovering cetaceans and telling them apart

» Meeting the real sirens of the sea

» Getting friendly with some pinnipeds

» Hanging out with polar bears and otter mammals

Chapter **15**

Getting Warm and Fuzzy with Marine Mammals

S ome marine mammals are obviously mammalian. When you look at a walrus, a seal, an otter, or a polar bear, you know it's a mammal. They're hairy and spend a lot of their time on land. Other marine mammals look like fish, swim like fish, and eat like fish, and they're certainly not hairy or even furry. If you weren't taught in grade school that dolphins and whales are mammals, you might still think they're fish. And unless they get washed up on a beach, they spend their entire lives in the water, like fish.

However, all these animals — whales, dolphins, walrus, sea lions, seals, otters, and polar bears — are mammals, just like you and us. Generally speaking, they're warm-blooded air-breathers, hairy or furry (yes, even whales and dolphins have a few whiskers on their chins at some point in their lives), most exhibit *heterodentition* (different teeth for different purposes), have unique bones in their middle ear, and the females give birth to live young and have mammary glands that produce milk to feed them. While there are always exceptions to these traits, the combination of them is what makes marine mammals, well, mammals. But what makes them so special?

In this chapter, we answer that question and go into a little more detail about the distinct groups of marine mammals — *cetaceans* (whales, dolphins, porpoises), *pinnipeds* (walrus, seals, sea lions), *sirenians* (manatees, dugong), and *marine fissipeds* (polar bears, otters).

What Makes Marine Mammals So Special?

Marine mammals, as the name suggests, are mammals that spend a significant amount of time in and around the ocean. They can be wholly aquatic (*obligate swimmers*), such as whales and dugong, or can spend extended periods of time on land, such as polar bears and some otters and seals.

Beyond being mammals, they all have special adaptations for ocean life. Most marine mammals have streamlined bodies to facilitate movement through water, and their limbs have been modified to facilitate swimming; for example, they have muscular (often flattened) tails, flippers, fins, and (in the case of the polar bear and otter) large flattened paws. A few additional adaptations enable them to stay warm and toasty, to breathe more easily when they surface, to catch and eat their favorite seafood, and to deal with all that salt in their water, as explained in the following sections.

Staying toasty

The ocean is a generally chilly environment, so marine mammals need adaptations to help regulate their body heat, and here they are:

>> **Fur:** Most marine mammals that spend part of their lives on land have thick, dense fur, and they dress in layers, with long, thick, interlocking (guard) hairs that seal out water, and finer, shorter, more densely packed hairs that form a dry, fluffy undercoat for warmth. (One of the drawbacks of this two-layer system is that it relies on trapped air, which discourages deep dives — air increases buoyancy and can get squeezed out by water pressure during dives.)

>> **Fat (blubber):** Marine mammals that spend their entire lives in water have little, if any fur, relying instead on a thick layer of fat (referred to as *blubber*) directly below the skin. In some marine mammals, primarily in cetaceans (whales, dolphins, and porpoises), this blubber layer can be incredibly thick: up to about 50 centimeters (20 inches) in bowhead whales. This blubber keeps 'em warm, even under pressure; it provides just enough buoyancy, so they can deep-dive for food; and it functions as a fuel reserve for long migrations.

>> **Countercurrent heat exchange:** Some marine mammals have a high-tech circulatory system for conserving heat. The arteries, which carry warm blood away from the heart are surrounded by veins that carry cooler blood back to the heart. This system maintains a more uniform temperature difference between the fluids, thereby reducing heat loss.

Breathing easy

Mammals can't breathe underwater. They need to surface for air. So, they must be able to store oxygen and use it efficiently. Several adaptations and best practices help marine mammals breathe easier and remain below the surface longer:

>> **Blowholes:** Whales, dolphins, and porpoises have their nostrils positioned at the top of their heads, so they can breathe more comfortably when they surface.

>> **Diving reflex:** All air-breathing vertebrates, including mammals, have a diving reflex (called *bradycardia*) that slows the heart rate and redistributes oxygen to the heart and brain to conserve it. The heart rate of Weddell seals while diving has been recorded at as low as four beats per minute (in contrast, a normal resting human heart rate is between 60 and 100 beats per minute).

>> **Reduced physical movement:** This may seem counterintuitive, but marine mammals often exhale to reduce buoyancy and then simply point their bodies downward to sink without using their limbs to propel themselves. They rely on oxygen stored in parts of their bodies other than their lungs to sustain them underwater.

>> **Greater total body oxygen stores:** Most diving mammals have adaptations that enable them to store oxygen throughout their bodies, such as larger spleens, higher red blood cell counts, and blood chemistry that allows for oxygen to be more densely packed into muscle tissue. The deepest divers among marine mammals rely very little on oxygen stored specifically in their lungs.

FUN FACT

Elephant seals can dive for up to two hours, swimming underwater for 2 kilometers (1.2 miles) chasing down food before they need to surface for air.

Adapting to their food source

In some ways, you can identify the different marine mammals just by inspecting their teeth and watching how and what they eat. Using this method, you can break marine mammals into four basic groups — crunchers, grazers, biters, and filter feeders.

Crunchers crunch

Sea otters are informally classified as "crunchers," feeding on hard-shelled organisms, such as mollusks, crustaceans, and sea urchins (as well as fish), all of which they're specially adapted to eat. They have blunt, flattened molars, short skulls, and strong jaws, which increase their bite force and enable them to crush or crack the shells of their prey. They're also special in that they collect prey with their paws instead of their jaws. Sea otters are extra special in that they're tool-users, using rocks to crack open thick shells.

FUN FACT

Cracks along the same side of different mussel shells opened by sea otters suggest that most otters are right-pawed.

Grazers graze

Sirenians (manatees and dugongs) feed almost exclusively on aquatic vegetation and are the only strict vegetarians among marine mammals. They graze primarily on seagrass and sea lettuce but also on algae and other marine and freshwater plants. They're equipped with strong molars to grind and chew plants along with a horny "crushing" plate for mashing their food.

Sirenians have dense bones that keep them submerged while they plod along the seafloor on their *pectoral* (front) fins, using their sense of smell or the sensitive bristles near their mouth to locate food. Like many mammalian herbivores (including horses and rabbits), sirenians are *hindgut fermenters,* meaning they have a really long intestine and a *caecum,* both of which are populated by symbiotic bacteria that help to digest tough *cellulose* (plant fiber).

Biters chomp, and some slurp

Most marine mammals are carnivorous, preying on fish, squid, mollusks, crustaceans, and other mammals. These predators all have at least some sharp teeth for biting their prey. Many whales and dolphins have exclusively sharp conical teeth; others, such as seals and polar bears, have canines, incisors, premolars, and molars.

In addition to their differences in dentition, they've developed a variety of clever methods for capturing their prey:

>> **Chasing and attacking:** This tried-and-true method is universal among the biters.

>> **Herding:** Dolphins often team up to herd fish into large bait balls and then drive them to the surface where they're easier to catch.

>> **Strand feeding:** Some species of dolphin corral fish into shallow water and deliberately rush the shore and beach themselves while catching fish that try to jump over them back into the water. Orcas also strand feed at times to catch seals.

>> **Suction feeding (slurping):** These "sucky" predators, which include sperm whales, walruses, and narwhals, get close to their prey and then generate suction by rapidly moving their tongue to the back of their mouth, effectively slurping up their next meal.

Male strap-toothed beaked whales (which look like dolphins) often have no other option than to suction feed. They have tusks that grow up from their bottom jaw and often curl over their upper jaw, so they can't open their mouths very wide to eat.

Filter feeders strain their food

Filter feeders have a unique way of gathering prey — they take a big gulp of water containing lots of small prey, usually krill or other zooplankton, close their mouths, force the water out between their teeth, and then either swallow their prey whole or chew and swallow.

The most famous filter feeders among marine mammals are baleen whales. Instead of teeth, they have close-fitting baleen plates growing down from their upper jaws like the teeth of a comb (see Figure 15-1). Each baleen plate consists of thousands of strands of keratin (the same substance in human hair and fingernails). Baleen whales also have pleated throats that can expand substantially as they fill with water (a blue whale's mouth can hold about 50,000 liters [13,200 gallons] of water when full).

Many whales use "bubble netting" to catch their food. They dive below a swarm of krill, blowing bubbles in a circular pattern on their way down so that the coil of bubbles tightens as it nears the surface. Krill feel confined by the bubbles, so the swarm becomes more concentrated, making it easier for the whale to gobble up more krill in a single gulp. Gray whales (another type of baleen whale) practice a combination of suction and filter feeding. They swim on their side near the seafloor, sucking up sediment and small crustaceans and then expelling the water and sediment through their baleen screen.

Crabeater, leopard, and Antarctic fur seals can also filter feed, but instead of baleen, they have designer teeth that look as though they've been carved by hand (see Figure 15-2).

FIGURE 15-1:
A humpback
whale feeding
with baleen
visible.

FIGURE 15-2:
Crabeater
seal baring
its teeth.

Adapting to salt water

"Water, water everywhere, and not a drop to drink" pretty much sums up the marine mammal's dilemma of how to quench their thirst when all they have to drink is seawater. How do they deal with this challenge? Two ways:

>> **Specialized kidneys:** Many marine mammals have kidneys that are more efficient than human kidneys at filtering out salt from water. If we were to drink a liter of salt water (not recommended, by the way), we would need to pee one and one-third liter of water to remove the salt. This would eventually lead to our dehydration and death. In contrast, some marine mammals produce urine that's up to two and a half times saltier than seawater, allowing them to reclaim much of the water they consume.

>> **A keto diet:** Many marine mammals get much of their water not by drinking it but through the metabolic process of breaking down fat, which releases water. Breaking down meat protein uses water, but breaking down fat releases water. The high-fat diets of many marine mammals, including polar bears and seals, provides them not only with energy, but also with the fresh water they need to survive.

Exhibiting special sensory adaptations

Marine mammal senses are similar to ours, though different senses are more or less developed than ours. They can see pretty well (or very well), hear extremely well, and feel vibrations and objects that touch their bodies, but their sense of smell is likely very weak or absent (with the exception of polar bears that have incredible noses), and they probably do not have a strong sense of taste. Of course, if you were eating raw fish and krill your entire life, you'd probably wish you had no sense of smell or taste. However, many marine mammals, especially those that spend their entire lives submerged, have special sensory adaptations, including the following:

>> **Larger, more spherical lenses:** While land mammals rely more on their corneas (the outer, clear, round structure that covers the iris and pupil) to focus their light, marine mammals rely more on their eyes' lenses, which provide more clarity underwater.

>> **Panoramic vision:** The eyes of cetaceans are positioned on either side of their heads (like fish), providing for a wider field of vision. However, they can move their eyes slightly forward and down giving them depth perception, which comes in handy for chasing prey.

- >> **Multifocus vision:** To see both out of water and underwater, some marine mammals, such as harbor seals, seem to have multifocal vision.

- >> **Larger pupils:** Some marine mammals that spend time on land, such as seals and sea lions, have irises that can contract to maximize the size of their pupils, enabling them to see in the low-light conditions when they dive.

- >> **Retroreflectors:** Like cats, polar bears, pinnipeds, and cetaceans have a layer of tissue (called a *tapetum lucidum*) located just behind the retina that reflects visible light back through the retina to improve sight in low-light conditions. They also have a higher number of light-receptor cells within their eyes.

- >> **Built-in goggles:** Some pinnipeds excrete a layer of mucus to protect their eyes from the salt, while others have a transparent eyelid (called a *nictitating membrane*) to protect their eyes.

- >> **Touch-sensitive whiskers:** Sirenians, sea otters, and pinnipeds have whiskers (called *vibrissae*), which are thought to be able to detect the movements of prey and predators. They can also move these whiskers back and forth to feel around for food.

- >> **Aquatic ears:** Marine mammals that spend a lot of time underwater, including pinnipeds and cetaceans, have no external *pinnae* (outer ear part). They're built more for underwater hearing and not getting water in their ear canals. Toothed whales are thought to pick up vibrations through a fatty area of their jawbone, known as the *acoustic window*, which is then transferred to the middle ear. Scientists aren't sure how baleen whales hear, but they think vibrations are transferred from their skulls to their ears. (Surely some clever inventor is fast at work developing the iSkull.)

- >> **Echolocation:** *Echolocation* involves emitting a sound, perceiving the sound as it's reflected back, and calculating an object's position based on the reflected sound; for example, the faster and more intense the reflected sound is, the closer and bigger the object is. Dolphins and bats are famous for their mastery of echolocation.

REMEMBER

Because sound is so important for cetaceans (for feeding, navigation, and communication), scientists are increasingly concerned about all the noise humans produce underwater. Sound travels faster and farther through water than air, so noises from boats, seismic testing, drilling, blasting, sonar, and more, can be very disruptive and even deadly to marine wildlife. (See Chapter 21 for more about the problem of sound pollution.)

Getting Acquainted with the Cetaceans: Whales, Dolphins, and Porpoises

Cetaceans (pronounced se-*tay*-shunz) are marine mammals that look and act more like fish than mammals. They include whales, dolphins, and porpoises. So, what's the difference? Technically speaking, they're all whales, but they're divided into two groups and several subgroups based mostly on their dental records:

» **Baleen whales** are 15 species of long whales with streamlined bodies, pleated throats which expand for feeding, a double blowhole, and baleen (instead of teeth). They included humpback whales, blue whales, minke whales, right whales, and grey whales.

» **Toothed whales** are 76 species of generally smaller whales characterized by teeth (instead of baleen) and a single blowhole. They include everyone's fav — dolphins — plus porpoises, beaked whales, sperm whales, narwhals, and beluga. Toothed whales are often broken down into several less formal subgroups:

  ● **Whales:** Large-toothed whales are called whales, even if they happen to look more like dolphins or porpoises. They include sperm whales, beaked whales, and bottlenose whales (different from the more well-known bottlenose dolphin).

  ● **Dolphins** include orcas (a.k.a. killer whales — confusing, we know), Risso's dolphins, snubfin dolphins, river dolphins, right whale dolphins, spinner dolphins, common dolphins, and bottlenose dolphins, to name a few. They vary considerably in shape, coloration, and size. Their *dorsal* (back) fin can be tall and pointy, curved, rounded, nublike, or completely absent; they may have a long beak or one that's not even discernible; and their bodies can be patterned in black, white, pink, grey, yellow, brown/beige, or cream, with large splashes of color, spots, or stripes.

  ● **Porpoises,** closely related to dolphins, account for only a handful of the toothed whales, including the harbor porpoise, vaquita (which are highly endangered), and spectacled porpoise. They're generally black, white, brown, or grey with a stocky body and a rounded head.

  ● ***Monodontidae*** (meaning "one tooth") comprise two species — beluga and narwhal — which are less than 6 meters (18 feet) long with stocky bodies, bulbous heads, broad rounded flippers, and no dorsal fin. The narwhal literally has one tooth sticking out of its head that looks more like a unicorn horn than a tooth, whereas the beluga has a complete set of 30 to 40 teeth.

Sizing up baleen whales

Baleen whales hold bragging rights for having among their membership the largest animal on the planet — the blue whale (see Figure 15-3), which can reach lengths of 33 meters (108 feet, which is longer than two school buses lined up end-to-end) and weigh as much as 136,000 kilograms (nearly 300,000 pounds). The tongue of a blue whale weighs as much as an elephant, and its heart is about 125 cubic feet (about five times the size of a hot tub). Some paleontologists believe that the blue whale is the largest animal on Earth . . . *ever!*

FIGURE 15-3:
A blue whale, the largest-ever animal on Earth.

Source: Cristina Mittermeier – www.sealegacy.org

Blue whales have been able to evolve into these massive creatures because their body mass is supported by salt water. A land animal the size of a blue whale would be crushed by its own weight. Oddly enough, the largest animal on the planet feeds on some of the smallest prey — krill. At certain times of the year, a single blue whale eats an estimated four tons of krill daily (talk about binging!).

Here are a few more baleen whales representing the group:

>> **Fin whale:** The second largest animal on the planet (and probably of all time), fin whales earned the nickname "greyhounds of the sea" due to their sleek bodies and their speed. You can identify fin whales by the white patch on only the right side of their bottom jaw.

>> **Sei whale:** Another speedy whale, the sei (pronounced "say") is the third largest in the group. The sei whale is unique in that when it dives it doesn't arch its back like other whales do. Instead, it simply sinks below the surface like a submarine.

>> **Humpback whale:** Named after the distinctive hump on their backs, these whales (see Figure 15-4) are known for their long migrations. Annually, they travel up to 16,000 miles between their feeding grounds in the Arctic or Antarctica to their summer breeding grounds nearer the equator. Growing up to about 15 meters (50 feet) long and up to 27,000 kilograms (60,000 pounds), humpbacks are considered medium-sized baleen whales, which gives you some idea of just how large the big baleens really are.

FIGURE 15-4: A mother and calf humpback whale in Hawaii. (Photo taken during research activities pursuant to NMFS research permit #10018 & 895-1450 while working with the Keiki Kohala Project.)

Source: Cesere Brothers – www.CesereBrothers.com

>> **Bryde's whale:** Named after Johan Bryde, a Norwegian who built the first whaling stations in South Africa (which doesn't seem right), Bryde's whale may be two or three species of baleen whale. They have three prominent ridges in front of their blowhole; sleek bodies; and slender, pointed flippers. They seem to prefer the temperate waters of the Atlantic, Indian, and Pacific oceans.

>> **Bowhead whale:** Named after their bow-shaped jaw, members of this group have the distinction of being the only whale native to the Arctic and subarctic waters. The bowhead can ram through ice 18 centimeters (7 inches) thick and leap almost entirely out of water, which is quite a feat for a whale that grows up to 18 meters (60 feet) long weighing up to 54,500 kilograms (120,000 pounds). The bowhead is also thought by some to be the longest living mammal on the planet, living up to 200 years.

In one bowhead whale, a harpoon was found that was made in the 1880s, suggesting that the whale was over 120 years old. Shame on whoever launched that harpoon! But then again, score one for the home team — that whale hunter is long gone, while that mighty bowhead lives on.

>> **Minke whales:** Minke (rhymes with pinkie) whales are two species — common and Antarctic. Among baleen whales, minkes are considered small, growing up to about 6 meters (18 feet) long and weighing in at a mere 5,400 kilograms (12,000 pounds). They're black on top and white on the bottom, and they tend to summer near the poles and spend their winters nearer to the equator.

>> **Right whale:** Except for the pygmy right whale, which grows only about 6 meters (20 feet) long and weighs only up to about 4,500 kilograms (10,000 pounds), right whales are fairly large. The North Pacific right whale can grow up to 15 meters (49 feet) and weigh as much as 81,600 kilograms (180,000 pounds). Right whales are often identified by the rough white patches of skin on their heads. As for their name, these unfortunate whales were singled out by whale hunters as the "right whales" to kill, because they're large, slow, and they float on the surface after they're killed.

>> **Gray whales:** Another species known for long migrations are the gray whales, which are, well, gray, dappled in white. Gray whales have a smaller head (proportionate to their body) compared to other baleen whales and a series of bumps on their back in place of a dorsal fin. They're also the only bottom feeders of the group. They grow up to about 14 meters (45 feet) long and weigh up to 36,000 kilograms (80,000 pounds).

Gray whales were historically referred to as "devil fish," because they were known to fight back against whaling ships by ramming boats or flipping over the smaller scouting vessels. Today, grays are known as gentle giants and are famous for "presenting" their babies to humans in tiny boats in Mexico — a bucket list experience for sure. As you can see in Figure 15-5, the mother is facing the boat, and the calf is facing away. Look closely, and you can see the calf's right eye looking up and the gentle curve of its mouth.

Source: Claudio Contreras-Koob – www.wildcoast.org

WHALES UNDER SIEGE

Unfortunately, many species of large whales were nearly hunted to extinction, and many are still threatened or endangered as a result of whaling (some of which is still going on — we're looking at you Japan, Norway, and Iceland). As long-lived species, whales take a while to reach maturity and reproduce. As larger, older whales were hunted, fewer adults remained to have calves, and juvenile animals were being taken before they could reproduce, so populations took a lot longer to recover. Thankfully, numbers of some species, including humpbacks, are increasing with some starting to return to pre-industrial populations.

However, whales now face additional threats, including increasing shipping traffic (and risk of being struck by boats), increasing marine noise, an increasing risk of becoming entangled in abandoned fishing gear and nets, and the pervasive threats caused by climate change. The north Atlantic right whale, for instance, was moved from "endangered" to "critically endangered" in July 2020, with only an estimated 400 individuals remaining. Before whaling, hundreds of thousands of these whales were thought to roam the Atlantic, filling the ocean with their songs, songs which have now largely gone silent.

Sinking your teeth into the toothed whales

Toothed whales include what are commonly known as whales along with dolphins and porpoises, but for organizational purposes, we decided to divide them into three groups — whales (sperm whales and beaked whales), dolphins, and porpoises. Generally, the toothed whales are smaller and faster than their baleen cousins, in addition to their dental differences.

Sperm whales and beaked whales

Toothed whales, including sperm and beaked whales, are distinguished from dolphins and porpoises mostly because of their size, even though many of them look more like dolphins or porpoises than what most people think of as whales. Members in this group are generally broken down into sperm whales and beaked whales:

>> **Sperm whales:** The largest of the toothed whales, the sperm whale can grow up to 18 meters (59 feet) long and weigh as much as 54,000 kilograms (119,000 pounds). They have dark, wrinkly skin, small jaws, and a large, boxy head that contains the largest brain in the animal kingdom (see Figure 15-6). In place of a dorsal fin, they have a round bump and "knuckles" toward their tails. They're well known for their deep dives and epic battles with whale hunters (like Ahab) and giant squid, using echolocation to hunt down their prey in the darkest depths.

FUN FACT

On the other end of the scale is the lightest whale, the dwarf sperm whale, which grows to only 2.6 meters (8.5 feet) and weighs only 136 to 272 kilograms (300 to 600 pounds).

>> **Beaked whales:** Beaked whales (which include bottlenose whales) are big like whales but have a nose (rostrum) like that of a dolphin or a porpoise. Their teeth stick out from their lower jaw, making them look as though they're in bad need of some dental work. Beaked whales hang out mostly offshore in deep waters, where they deep-dive for prey, as highlighted in the nearby sidebar. They range in length from about 5 to 13 meters (16 to 42 feet), and they're rarely spotted at the surface.

>> **Killer whales:** Fooled you! These are really dolphins . . . still killers, but dolphins, not whales (see the next section on dolphins).

Source: Keith Ellenbogen – www.keithellenbogen.com

FIGURE 15-6:
A sperm whale.

NATURAL-BORN DIVERS

The record for longest marine mammal dive is held by the Cuvier's beaked whale — diving for 138 minutes. Coincidentally, this rare zombie-looking animal also holds the record for the deepest dive at 2,992 meters (9,816 feet) — eight times deeper than the Empire State Building is tall.

At 1,000 meters (3,280 feet) deep, they experience 100 times surface pressure. While diving, their lungs collapse, so they rely on oxygen stored in their blood and muscles, making them less susceptible to the bends. (The *bends* is a potentially deadly condition that human divers can suffer as well when they surface too quickly after a dive — see Chapter 18 for details.)

Whales aren't completely immune to the bends, however; some beaked whales that washed ashore in the Canary Islands were found to have the bends. Experts think that this incident may have been caused by sonar exercises nearby — trying to avoid the noise, these whales may have ascended too quickly.

Deep-diving whales also have a rib cage that can fold inward to reduce air pockets within their bodies, and little indentations on the sides of their bodies for their pectoral fins — so they can be tucked away to reduce drag.

Dolphins

To anyone who's spent any time in, on, or around the ocean, the dolphin is a familiar sight. It has a characteristically long, sleek body, an extended snout, and a mouth packed with more than 200 tiny, cone-shaped teeth. Like other cetaceans, dolphins have powerful tails to propel them through the water. They can maintain a cruising speed of about 30 kilometers (20 miles) per hour with bursts in excess of 40 kilometers (25 miles) per hour. They typically swim near the surface, coming up for air every 2 or 3 minutes, but are capable of diving to depths of 300 meters (1,000 feet) and staying underwater for more than 10 minutes. Their lungs are specially adapted to cope with the dramatic pressure changes they experience by descending and ascending so quickly.

More than 30 species of dolphins cruise the world's oceans, and several freshwater species are commonly found in estuaries and far up into rivers. Dolphins typically eat the equivalent of about one third of their body weight daily and dine on a diet consisting primarily of fish, shrimp, and squid. They generally live in pods or even superpods (containing more than 1,000 individuals), and they typically hunt in small groups, working like a pack of wolves to herd smal fish before attacking the school.

At the age of 5 to 12 years for females and slightly later for males, dolphins reach sexual maturity. They typically mate in the spring and give birth approximately one year later. Most females bear only one baby at a time. The baby dolphin begins swimming immediately and stays with the mother for about 18 months to nurse.

Here are a few representative dolphin species:

>> **Common bottlenose dolphins** (see Figure 15-7) live all around the world. Some species prefer coastal living, while others spend more time in the open ocean. They can be very acrobatic, leaping high into the air.

>> **Spinner dolphins** are known for their acrobatics, sometimes spinning after becoming airborne (show-offs!).

>> **Short-beaked oceanic dolphins** such as the hourglass dolphin, the dusky dolphin, or the white-sided dolphin have much shorter rostrums, making them look more like porpoises.

>> **Right whale dolphins** are characterized by their lack of a dorsal fin.

>> **Risso's dolphins** are large, stocky dolphins with blunt heads and no discernable beak and are often recognized by their many scars and scratches from squid and from fights with other Risso's dolphins.

FIGURE 15-7:
A pair of
common
bottlenose
dolphins.

Source: Romona Robbins – www.RomonaRobbins.com

>> **Orcas** are the largest of the dolphins with the largest recorded being about
10 meters (33 feet) long and weighing up to about 5,400 kilograms (12,000
pounds); see Figure 15-8. Their distinctive coloration, white eye patch, and tall
dorsal fin make them an instantly recognizable species. Commonly called "killer
whale," orcas are highly effective hunters, with some preying on seals and
dolphins, and even working together to take down whales. (None on record
has ever hurt a person in the wild — we repeat, *in the wild,* where they belong.)

Porpoises

Telling the difference between a dolphin and a porpoise can be quite a challenge. Generally, dolphins have longer noses (beak or rostrum), more streamlined bodies, curvier dorsal fins, and conical teeth. Porpoises typically have flatter,
spade-shaped teeth, more rounded bodies, more triangular dorsal fins, and no
discernible beak. Dolphins are also thought to be more vocal and are generally
larger than porpoises.

Porpoises account for only seven cetacean species — the harbor porpoise, vaquita
(or Gulf of California harbor porpoise, which are on the verge of extinction), Dall's
porpoise, Burmeister's porpoise, spectacled porpoise, Indo-Pacific finless porpoise, and narrow-ridged finless porpoise.

FIGURE 15-8:
An orca.

Photo by Christopher Michel. Licensed under CC BY 4.0

Narwhals and belugas

Narwhals and belugas comprise their own toothed whale family. It's hard to say which is weirder. The narwhal has a single tooth that's 3 meters (10 feet) long protruding from its upper lip (see Figure 15-9), whereas the beluga looks like a pale alien from another planet whose brain is too big for its skull (see Figure 15-10). Neither has a dorsal fin — only a ridge where the fin would be. Both have a small head and a short beak, and their neck bones aren't fused, so they can turn their heads without turning their entire bodies (which would make them good drivers, if they were to drive a car).

If that's not weird enough for you, scientists recently discovered a cross between a narwhal and a beluga, which they're calling a *narluga*. That's right, the two weirdest whales in the ocean managed to hook up and have a baby.

Swimming with the Sirenians: Manatees and Dugongs

Sirenian (pronounced sigh-*reen*-ee-uhn) is a word from Greek mythology referring to creatures that were half bird, half woman who sang so seductively that sailors would crash their ships on the rocky shores in pursuit of them. However, looking at a manatee or a dugong, the first word that pops into your head probably isn't "siren" or "seductive," unless, of course, you're a manatee or a dugong. Sea cow is a more apt description. They're large and portly like cows, they lumber like cows, and they graze sort of like cows, though they're actually more closely related to elephants. Unfortunately, it is that slow lumbering that makes them vulnerable to impacts like boat strikes, leaving a lot of baby manatees orphaned (see Figure 15-11).

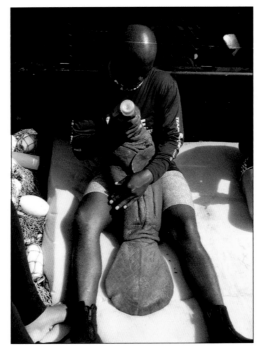

FIGURE 15-11: Our friend Jamal Galves feeding a rescued orphaned baby manatee.

Source: Jamal Galves – @therealmanateeman, Clearwater Marine Aquarium

Four species of sirenians now graze in tropical and semi-tropical coastal marine and freshwater environments — the West Indian manatee, the African manatee, the Amazonian manatee, and the dugong (strictly marine). The Steller's sea cow, now extinct, lived in Arctic waters near Alaska and off the coast of Russia and

Japan, feeding on kelp. Sadly, only a few decades after Europeans stumbled upon them, they were hunted to extinction around 1768, as a valuable source of fat and meat (so much for sustainable harvests).

REMEMBER

Manatees and dugong differ in the following ways:

>> **Environment:** Dugong live only in marine environments, whereas manatees also hang out in estuaries or in freshwater rivers or lakes.

>> **Tail shape:** Dugongs have fluked tails like whales, whereas manatee tails are shaped like roundish canoe paddles.

>> **Size:** Manatees are generally larger, growing up to 4 meters (13 feet) long and weighing up to about 1,600 kilograms (3,500 pounds). On average, dugong grow up to about 3 meters (10 feet) long and weigh up to 500 kilograms (1,000 pounds).

>> **Face:** Manatees have a rounder, more forward-pointing face. A dugong's face points more downward, and their mouth is more like a short, wide elephant trunk. They're both adorable but probably very sloppy kissers.

>> **Life expectancy:** Dugongs live about 70 years — twice as long as manatees.

While both manatees and dugongs are generally solitary or live in small groups, they can form large gatherings, such as those near warm-water outflows at power stations in Florida during the winter. While manatees and dugongs are generally slow animals (slow enough that algae can grow on their backs), they can show quick bursts of speed. Both manatees and dugongs are threatened by poaching, habitat degradation, entanglement in fishing gear, and boat strikes. Please, please drive slowly when boating in manatee and dugong waters — a single hit with a boat propeller can kill or seriously injure one of these gentle giants and leave their offspring orphaned. Despite these threats, the tireless work of people like Jamal Galves, who dedicate their lives to educating communities about and rescuing and rehabilitating these animals, there is hope.

Seals, Walruses, and Other Pinnipeds

Pinnipeds (meaning "fin-footed") include seals, sea lions, and walruses, all of which are thick-furred, big-eyed, semi-aquatic carnivorous animals that feed in the water and breed onshore. During breeding season, pinnipeds are known to form large colonies; for example, the Cape fur seal colony in Cape Cross, Namibia is thought to be composed of 80,000 to 100,000 fur seals! While most pinniped species are marine, the Baikal seal lives exclusively in freshwater habitats, such as Lake Baikal in Russia.

An interesting fact of pinniped biology is their strategy of *delayed implantation* — female pinnipeds can put the development of their offspring on hold while they're still embryos to time the birth to match more favorable environmental conditions.

Pinnipeds can be divided into three groups: earless (true) seals, eared seals (such as fur seals and sea lions), and walruses.

Earless (true) seals

The earless seals (true seals) have smaller flippers than other pinnipeds and prefer to lounge around on the beach rather than sit upright. Unlike walruses and eared seals, they move around on land somewhat like a caterpillar, earning them the nickname "crawling seals." In the water, however, they're incredibly graceful, using their back flippers to propel themselves forward. Well-known members of this group include the leopard seal, crabeater seal, and Weddell seal (see Figure 15-12) from the Southern Hemisphere; the grey seal, ringed seal, harp seal, and harbor seal in the Northern Hemisphere; and the northern and southern elephant seals from their respective polar regions.

FIGURE 15-12:
Ashlan in Antarctica with Weddell seals in the background.

Source: Ashlan Cousteau

While everyone has heard of polar bears, the apex predator in the Arctic, not as many people are familiar with leopard seals, the apex predator in the Antarctic (remember, polar bears are only in the Arctic). Leopard seals (see Figure 15-13) are a fearsome predator. They can get up to 3 meters (11 feet) long and weigh up to 380 kilograms

(840 pounds)! They are called leopard seals because they have a grey coat with black dots similar to a leopard's spots. They are also the only seal to feed on warm-blooded animals — namely, other seals and penguins — though squid and shellfish are also an important part of their diet.

FIGURE 15-13:
The leopard
seal, apex
predator of
Antarctica.

Source: Cristina Mittermeier – www.sealegacy.org

FUN FACT

The southern elephant seal is the largest pinniped, with some males weighing in excess of 3,700 kilograms (8,000 pounds) and growing up to 6 meters (roughly 20 feet) long. Elephant seals are named for the "trunks" the males have, which they inflate and use to produce loud noises to scare off rival males and attract females. If these roars aren't sufficient to drive away competing suitors, males can become engaged in vicious, bloody battles to become beachmaster and control a harem of females. Yeah, don't mess with these bad boys.

Eared seals

Eared seals are unique among pinnipeds in that they have ear flaps rather than the absence of external ears that characterizes other pinnipeds. They can also rotate both their front and back limbs, allowing them to walk on land (if you can call what they do walking) and use their front flippers to propel themselves and their rear flippers to steer underwater.

Members of this group include fur seals and sea lions, the main difference being their fur — fur seals have longer fur with a softer, fluffier undercoating absent in sea lions. Sea lion species are also generally a bit larger than fur seals and have smaller hind flippers. In the water, both are quick, agile, and playful — their antics underwater support their nickname "dogs of the sea." One of our favorite dives was with young sea lions, who were constantly trying to steal our fins and nibble on our heads (see Figure 15-14).

FIGURE 15-14: Sea lions in British Columbia playfully biting Ashlan's head during a dive. (Don't worry, they're gentle.)

Source: Philippe Cousteau

Walruses

You can easily spot the walruses in a crowd of pinnipeds (see Figure 15-15) — they're the large ones with the wrinkly brown skin, bushy moustaches, and long tusks (both males and females have them). Biologists initially thought walruses only used their tusks to dig up their favorite food — shellfish (although they also eat fish, crustaceans, and other invertebrates). Now they believe that walruses also use their tusks for maintaining breathing holes in the ice, hauling themselves out onto the ice, and fighting (in the case of males).

Walruses spend most of their lives in the Arctic, where they rely on their thick layers of blubber to keep warm in freezing cold water. Like eared seals, they're able to rotate all four flippers to walk on land, but their swimming style is more like that of true seals. Walruses are some of the largest pinnipeds, second only to elephant seals. Male walruses can weigh up to 2,000 kilograms (4,400 pounds).

FIGURE 15-15:
A walrus on
sea ice.

Stepping Out with the Marine Fissipeds

Fissipeds are carnivorous mammals that have separated toes, such as dogs, cats, bears, and racoons. Marine fissipeds are those that live in marine environments, namely polar bears and otters.

The ever-popular polar bears

Many people probably don't think of polar bears as *marine* mammals; they think of them as bears that live in the Arctic and drink Coca-Cola. However, polar bears can swim long distances for days at a time, they depend on the ocean for food, and they spend most of their time on sea ice. They rely on many adaptations to thrive specifically in and around icy Arctic waters, including paws that function as both paddles and snowshoes, dermal bumps on the pads of their paws (like studs on track shoes) for improving traction on the ice, specialized kidneys to help filter out salt, thick layers of fat and fur to stay warm, hollow white fur to blend into the ice and snow (hollow to trap air that insulates them), and black skin under that white fur to better absorb heat from sun light. See Figure 15-16.

Polar bears are the largest of all bears and are apex predators in the Arctic ecosystem. Their preferred food is seals, but they're not picky eaters. They'll eat pretty much anything — fish, birds, reindeer, muskox, crustaceans, bird eggs, or even food items discarded by other animals including people. They have a strong sense of smell and can detect food up to a kilometer (about two-thirds of a mile) away.

FIGURE 15-16: A solitary polar bear on sea ice.

Source: Cristina Mittermeier – www.sealegacy.org

POLAR BEARS AT RISK

Polar bears may rule the Arctic, but they face serious environmental threats. As apex predators, they're affected by *biomagnification* — the process by which toxins increase their concentration in organisms as they're passed to higher and higher levels in a food chain. Toxins from oil spills and oil clean-ups can enter a food chain, and polar bears may also ingest these toxins as they swim or when they clean themselves. Oil can also degrade the insulative property of their fur, increasing the risk of hypothermia.

Climate change is also a huge problem for polar bears, because the ice they rely on is melting. Polar bears hunt seals on this ice, and as it melts earlier in the year, it's not thick enough to support their weight when food is scarce. As a result, more polar bears are venturing closer to human settlements for food or having to swim farther to find food, and females struggle to find suitable places for their dens.

Climate change activists commonly use photos of starving polar bears to drive home the message of just how tragic globally warming can be to the wildlife that have no control over what we humans decide to do.

During the fall/autumn, females dig maternity dens in the snow, where they enter a resting state for the winter, similar to but not true hibernation. Within these dens, they give birth to and then nurse their cubs until the weather warms. They may reuse the same den for several years.

Otters you "oughter" know

Otters are 13 species of mostly freshwater mammals. Only two species live in the ocean — sea otters and marine otters. They are part of the mustelid family that includes weasels, ferrets, badgers and wolverines.

Sea otters live in the northern Pacific Ocean, near rocky coastlines and kelp forests. They're the heaviest among otters. Males can grow up to 1.5 meters (about 5 feet) and weigh up to about 45 kilograms (100 pounds). They have a thick neck; a large, round head; a blunt snout; short front limbs with retractable claws for grasping; and back legs that are more like flippers. Under each front leg is a pouch the otter uses to collect food for diving or to store a favorite rock for smashing open shellfish; okay, that is smart. Sea otters can also close their ears and noses while diving. As they age, they develop their iconic blond heads.

Sea otters seem to have a leisurely life (see Figure 15-17) and are known for their resting behavior — lying on their backs in groups and sometimes wrapping themselves in kelp to keep from floating away. After eating, they groom and wash themselves extensively and blow air into their fur to help keep themselves warm.

Marine otters (also known as sea cats) the smaller cousin to sea otters, are the smallest of the marine mammals. They live in coastal regions, along the rough, rocky west coasts of South America. They spend far more time on land than sea otters do, entering the water only to feed, and they may spend some time in freshwater rivers. They have a longer, more tapered tail, and are a bit slimmer with a longer face and whiskers. Their teeth are more adapted for slicing than crushing, and they feed mostly on crustaceans, mollusks, fish, worms, and echinoderms, as well as the occasional bird or terrestrial animal. They're usually seen alone or in small groups and give birth in dens on land.

FUN FACT

Sea otters have the most hair of any mammal — up to 1 million hairs per square inch! By comparison, humans have about 800 to 1,200 hairs per square inch, depending on ethnicity and hair color.

FIGURE 15-17:
A sea otter
preening
its fur.

Source: Paul Nicklen – www.sealegacy.org

MODERN THREATS TO OTTERS

While once thought to number around 150,000 to 300,000 individuals, sea otters were nearly hunted to extinction for their soft fur. In 1911, their population was estimated at 1,000 to 2,000. Thankfully, populations in some areas are rebounding, with some otters being relocated to assist with recolonization.

Today, sea and marine otters are threatened more by indirect human activity, such as pollution and boat strikes. In particular they suffer terribly when oil spills occur. The oil can soak their coats, leading to hypothermia, and they can accidently ingest the oil while swimming or cleaning themselves, which can result in liver damage.

Sea otters are also susceptible to *toxoplasmosis* — a disease caused by a parasite found in cat feces. The moral of this story — dispose of cat litter properly. *Don't* flush it down the toilet; the parasite's *oocysts* (egglike structures) can make their way through our water systems and into the ocean to the ocean and cause potentially fatal infections in these adorable and important mammals.

4

Grasping Basic Ocean Physics

Understand the various ways in which ocean water moves — waves, tides, currents, upwellings, and more.

Trace the path of the ocean's *global conveyor belt* — the system of interacting ocean currents responsible for transporting water, heat, and nutrients around the world.

Appreciate the crucial role the ocean plays in modulating climate and producing whether systems over both land and sea.

Get to know what a tropical cyclone is, how it forms, and how it grows in size and intensity.

Answer the question, "What's the difference between a hurricane, a typhoon, and a cyclone?" (Clue: It's sort of a trick question.)

IN THIS CHAPTER

» **Understanding how waves, tides, and tidal waves are formed**

» **Mixing things up with upwelling and downwelling**

» **Going with the flow . . . of ocean currents**

» **Grasping the fundamentals of gyres and trade winds**

» **Recognizing the forces that control sea levels**

Chapter **16**

Following the Ocean in Motion

When you look out across the ocean on a calm day it appears still, but that perceived stillness is a deception. Ocean water is in constant motion, pushed and pulled by gravitational forces, whipped by winds, shaken by earthquakes, stirred by storms, and churned by variations in water density (determined by its temperature and salinity). All this motion is essential for mixing and distributing nutrients, modulating global climate, and influencing weather patterns. The currents that result from these large movements of water also impact ocean navigation while providing potential sources of green energy.

In this chapter, we examine the different ways ocean water flows and the various mechanisms that drive the flow of water and energy in the ocean.

Meeting the World's Largest Wave Machine

Most water parks these days have a wave pool that produces artificial waves. Not to ruin it for you, but these wave machines typically function like oversized toilet tanks — collecting water, then dumping it all at once into one end of the pool. The sudden displacement of water sends a wave that travels to the opposite end of the pool, knocking down kids like bowling pins.

Compared to the ocean's capacity for generating waves, even the largest wave machines are rinky-dink operations. While artificial waves max out at about 3 meters (10 feet) high, large ocean waves are in the range of about 24 meters (80 feet) high. These natural waves come in four forms — surface waves, tsunamis, tidal waves, and internal waves.

Plain ol' surface waves

Surface waves, also called *wind-driven waves,* are just that. As wind blows across the surface of the water, it pushes the water up into a wave. As the water falls back down under its own weight, it pushes up the water in front of it. As more wind energy is added, the waves become larger. While the wave moves in the direction of the wind, the water stays put for the most part; it moves up and down and in a circular pattern that remains stationary below the surface, as shown in Figure 16-1.

WHAT IS A WAVE, ANYWAY?

Do you love to play in the waves, surf, or fall asleep to the calming rhythm of waves washing up on shore? Well then, you love energy. While waves look like water in motion, they're actually energy in motion. The water pretty much stays where it is, rising and falling as a wave of energy moves through it.

Think about when you swim out beyond where the waves are breaking, and you just rise and fall with the waves. You don't actually move closer to or farther from the beach (unless there's an undertow, which is something quite different). You just kind of stay in one place and bob up and down.

But this gentle wave motion is deceptive. Ocean waves are storehouses of considerable amounts of energy. By some estimates, just 2 percent of the energy in ocean waves would be enough to power the entire world!

As waves near the shore, the friction from the rising seafloor slows them down, and, like a slowdown on the freeway, the waves start to bunch up and squeeze together. As a result of the rising seafloor, they also get taller — the water has to go somewhere, and since it can't go down, it goes up. At the same time, the friction from the seafloor slows the bottom of the wave more than the top. Eventually, the top of the wave passes the bottom of it, causing the wave to curl forward and break (which is pretty gnarly, dude). The same thing happens when you run from shore into the water. The water slows your legs, but your upper body continues to move forward unhindered until you do a face-plant.

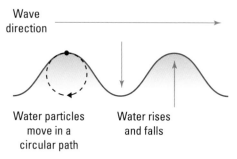

Wave direction

Water particles move in a circular path

Water rises and falls

© John Wiley & Sons, Inc.

FIGURE 16-1: Wave motion versus water motion.

How the waves break depends on the contour of the seafloor (yeah, they do, brah, just check out Figure 16-2):

>> **Spilling breakers** are the soft waves that break on relatively flat coastlines. Literally they just spill over onto the beach. Surfers call these "mushy waves."

>> **Plunging breakers** (which surfers like) are waves that break over steep shorelines or an abrupt change in the bottom, such as a reef. Because of the incline, the wave and its energy push up and over, plunging the wave in front of itself in a circular motion.

>> **Surging breakers** occur on very steep beaches. Since the bottom of the wave hits the ocean floor first, the wave is pushed up from the bottom. These breakers don't look scary and can seem not to break at all, but the outgoing wave that bounces back can be extremely powerful and dangerous (and can drag you quickly down and out).

>> **Collapsing breakers** are like plunging breakers that don't curl at the top. They just get tall then fall, usually producing some foam.

Tsunamis: So-called tidal waves

Tsunami is a Japanese word that means "harbor wave." It's a massive wave (usually a series of waves) caused by underwater disturbances such as earthquakes, landslides (underwater or coastal), or volcanic eruptions that displace huge volumes of water. What they're not caused by are tides, so they're not the same as tidal waves (which we describe in the next section).

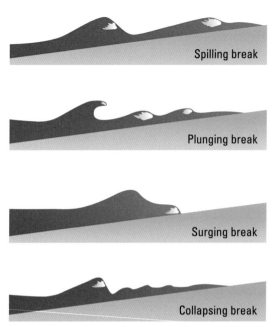

FIGURE 16-2: Types of breakers.

Depending on its size, where it hits, and how long it lasts, a tsunami can be devastating to property and lethal to any living thing that's in its path. If you're in a boat out at sea, you may not be able to tell the difference between a tsunami and a surface wave, but as a tsunami approaches the shoreline, the rising seafloor forces all that energy upwards, causing the tsunami waves to get high . . . like really high; in fact, some waves can reach the height of a ten-story building (about 30 meters or 100 feet tall). But it's not only the height that matters; the water is also moving at a fast clip. In the open ocean, a tsunami can travel faster than 800 kilometers (500 miles) per hour, slowing to about 32 to 64 kilometers (20 to 40 miles) per hour when it reaches shore. Tsunamis can push inland 300 meters (1,000 feet) or more (sometimes several kilometers or miles).

The Boxing Day Tsunami of 2004 was particularly devastating. A 9.1-magnitude earthquake in the Indian Ocean sent a force of energy that resulted in a massive surge of water that struck coastal areas of Sumatra, Thailand, India, and Sri Lanka, flattening buildings and killing more than 230,000 people in a matter of hours. The wave that struck Banda Aceh, on the northern tip of Sumatra, was the tallest one recorded from the quake at about 30 meters (nearly 100 feet) tall.

In 2011, a magnitude-9 earthquake in northeastern Japan triggered a tsunami with waves estimated to be as high as 40.5 meters (133 feet), flooding more than 200 square miles of coastal land. The tsunami killed 20,000 people, forced another 500,000 to evacuate, and triggered a nuclear emergency that Japan is still struggling with to this day.

While the waves coming ashore are certainly devasting, the rush of the water returning to the ocean can be just as, or even more, damaging (as the water drags back into the ocean everything it smashed, like trees, cars, buildings and more, and hitting anything that is still left standing with all this debris). Thankfully, many sensors are positioned all around the world to monitor underwater activity and predict potential tsunamis, so authorities can attempt to warn people early enough to get out of harm's way.

If you're on the coast and all of a sudden the water quickly and *extremely* recedes into to sea, don't scratch your head wondering what's going on. Head to high ground as fast as you can. This is called "drawback" and most likely means a tsunami is coming. The drawback may give you only a few seconds or a few minutes to run, but that could be the difference between life and death.

Rising and falling with the tides

Tidal waves sound big and scary, because "tidal wave" is frequently misused in reference to tsunamis. Unlike tsunamis, tidal waves are regularly reoccurring waves that are caused by *tides* — the daily rising and falling of sea level along coastal areas on different parts of the globe. Tides move so much water that they can cause the disappearance and reappearance of entire beaches and sandbars in a matter of hours. In most areas, the water level changes only about 1 to 3 meters (3 to 9 feet), but in some places, the level can change as much as 15 meters (50 feet) between lunch and dinner!

Tides are caused by the interaction of gravitational fields of three cosmic bodies — Earth, the sun, and the moon — along with the spinning of Earth, not by wind or geological events.

Back in the 1600s, Sir Isaac Newton explained tidal forces as due to a subtle imbalance between localized gravitational and centrifugal forces. Here's how it works: The *Earth-Moon system* revolves around its common *center of mass* — the point at which an imaginary rod connecting Earth and the moon would balance if placed on a fulcrum (see Figure 16-3).

Gravitational forces pull Earth and the moon toward one another, while the centrifugal force from spinning around the center of mass pushes them away from one another (like loving and hating someone at the same time). These two forces are balanced, keeping Earth and the moon from crashing into one another or flying apart.

However, at any given location on Earth's surface, gravitational and centrifugal forces may not be in perfect balance. The magnitude of the gravitational force at Earth's surface depends on the distance to the moon's center of gravity. The direction of the gravitational force is toward the moon's center of mass (see Figure 16-4).

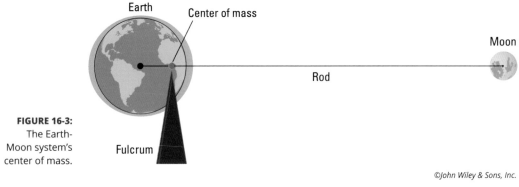

FIGURE 16-3:
The Earth-
Moon system's
center of mass.

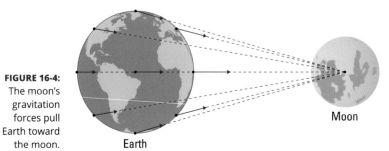

FIGURE 16-4:
The moon's
gravitation
forces pull
Earth toward
the moon.

Note that the solid arrows pointing from Earth toward the moon are longer on the side of Earth nearest the moon. The gravitational pull of the moon is strongest at locations nearest the moon and weakest at locations farthest from the moon.

In contrast, the centrifugal force is the same everywhere on Earth — it doesn't vary based on the distance to the moon's center of gravity, and it always points away from the moon and parallel to the imaginary line that connects the center of Earth and the center of the moon (see Figure 16-5).

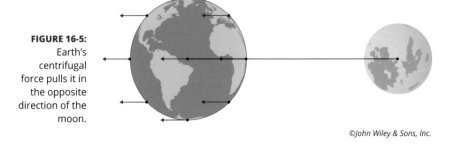

FIGURE 16-5:
Earth's
centrifugal
force pulls it in
the opposite
direction of the
moon.

Local differences in these two forces give rise to tide-generating forces that pull fluid water on both sides of Earth from the poles toward the imaginary line connecting the center of Earth and the center of the moon (see Figure 16-6).

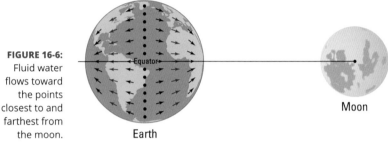

FIGURE 16-6: Fluid water flows toward the points closest to and farthest from the moon.

This water movement is what gives rise to the two tidal bulges (see Figure 16-7). The bulge on the side of Earth closest to the moon is slightly higher because the pull of the moon's gravity minus the offsetting push of the centrifugal force is greater on the side of Earth that faces the moon.

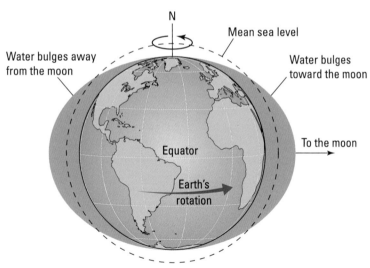

FIGURE 16-7: Water bulges on the side closest to and farthest from the moon.

REMEMBER

Newton's equilibrium theory does an elegant job of explaining the semidiurnal tide, but it assumes an Earth with no continents and ignores friction between the water and ocean sediments. The dynamic theory of tides allows for continents and friction and can explain why some tides are diurnal and some areas have no tides, but we'll leave that topic for another day (you're welcome). At this point, our brains feel as though they're bulging out the sides of our skulls.

When you factor in the sun, tides get more complex (see Figure 16-8). When the sun and moon are aligned, they each contribute their tugging power to create even higher high tides and lower low tides (called *spring tides*). When the sun and the moon are at right angles to each other in relation to Earth, their forces work against each other, resulting in less tidal variation around the globe (called *neap tides*).

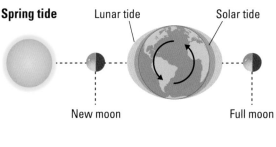

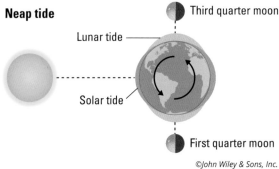

FIGURE 16-8:
Spring tide and neap tide.

The difference between the water level at high tide and low tide is called the *tidal range*. These ranges are generally stable and predictable unless a meteorological or geological event throws them off.

Going unnoticed: Internal waves

Internal waves occur below the surface when the interface between layers of different water densities (determined by temperature and salinity) is disturbed. Disturbances may be caused by the lower, denser layer of water being forced by tidal action over an obstacle on the seafloor, such as a ridge.

Internal waves can be much larger than surface waves because it takes a lot less energy to lift a mass of water into a less dense mass of water than it takes to lift that same mass of water into the air. While tsunamis generally max out at about 30 meters (100 feet) tall when they approach land, these underwater waves have a typical amplitude of 50 to 60 meters (about 160 to 200 feet) with some amplitudes in excess of 200 meters (about 660 feet)! Yet, they're rarely seen, often appearing on the surface as only a rough patch of water.

Though rarely observed except by people who study them, these internal waves can propagate horizontally and vertically, playing a key role mixing water in the ocean.

Upwelling and Downwelling in the Water Column

Currents, winds, and other forces create upwelling and downwelling. *Upwelling* occurs when cold, nutrient-rich water from the deep ocean is swept up into the shallow portion of the water column. This nutrient-rich water fuels phytoplankton growth, which is the main food source for zooplankton and small fish.

One place where upwelling occurs is where winds push water away from a shoreline, causing water from the bottom to rise up to replace it (see Figure 16-9). Upwelling also occurs in the open ocean when surface currents diverge. Once again, the deep water flows upward to replace the waters that are flowing away from each other. Because of all the nutrients they bring, areas where upwelling occurs are home to some of the most productive marine ecosystems. Although these areas cover only about 1 percent of the total area of the world's oceans, they account for nearly 50 percent of the fish harvested for human consumption.

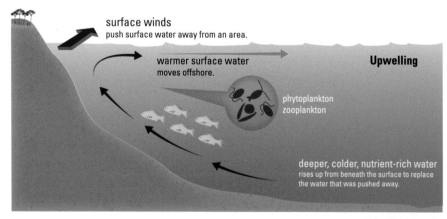

surface winds
push surface water away from an area.

warmer surface water
moves offshore.

Upwelling

phytoplankton
zooplankton

deeper, colder, nutrient-rich water
rises up from beneath the surface to replace
the water that was pushed away.

FIGURE 16-9:
Upwelling.

©John Wiley & Sons, Inc.

Downwelling is the opposite, occurring when surface waters are pushed down. Downwelling occurs where two ocean surface currents converge or water is pushed toward land. Downwelling also occurs off the coasts of Greenland and Antarctica when surface water gets denser (heavier) than the water below due to cooling and to increased salinity from sea ice formation. Downwelling leads to reduced productivity because surface waters are usually depleted of nutrients. However, they're just as important for keeping the ocean in motion.

Riding the Currents: The Ocean's Global Conveyor Belt

While waves are mostly energy moving through water, currents are continuous movements of water (no lazy water here). Think of currents as rivers flowing through the ocean.

Three mechanisms fuel currents:

>> **Tides:** The rising and falling of water levels create tidal currents along the coastline.

>> **Wind:** In addition to creating surface waves, wind drives currents near the ocean's surface, which can draw water up from the bottom (upwelling) or push water down to the bottom (downwelling).

>> **Thermohaline circulation:** Currents in surface and deep ocean water are created by variations in water density (due to both temperature and salinity). These currents also contribute to upwelling and downwelling.

You've already seen in the previous section how tides and wind move water, but how the heck does this thermohaline circulation thing work? First, keep the following facts in mind:

>> Seawater is denser (heavier) than fresh water because of its salinity, so it tends to sink.

>> Cold seawater is denser (heavier) than warm seawater, so it sinks.

>> Therefore, cold *and* salty seawater is denser than warmer, less salty seawater, so it sinks.

>> The North Pole and South Pole are cold enough to freeze salt water, which doesn't freeze until −1.5 degrees Celsius or 29 degrees Fahrenheit, unlike fresh water that freezes at 0 degrees Celsius or 32 degrees Fahrenheit.

Now, imagine you're off the coast of Greenland, where the water is very, very cold and dense. When water freezes, it leaves behind the salt, so as sea ice forms it sheds its salt into the water below, making that cold water even denser (heavier) due to the higher concentration of salt. This super dense water sinks, pushing its way down, like concert goers trying to get closer to the stage, which creates a downward current (downwelling). This really cold, salty, dense water (called "North Atlantic Deep Water" or NADW) dives down until it hits the seafloor, where it then flows southward along the bottom of the ocean between continents until it approaches Antarctica.

The surface water in the Southern Ocean, off the coast of Antarctica, is also very, very cold. (In fact, it's even colder than the water off the coast of Greenland.) This water also gets saltier due to sea ice formation. The resulting water (called "Antarctic Bottom Water" or AABW) sinks to the bottom (downwelling again) until it hits the seafloor, where it then flows northward.

From deep in the Southern Ocean, the mishmash of NADW and AABW flows in the deep toward the Indian and Pacific oceans. As the water flows northward, it begins to gradually receive warmth from above. Eventually, it becomes warm enough and light enough to upwell to the surface. Some of the deep water will upwell in the Indian Ocean. But some of the water will remain in the deep until it reaches the Pacific Ocean. Once the water reaches the surface, it will eventually flow back to where the journey began (the North Atlantic) and where it can start its trip all over again. This complete circuit is known as the *global conveyor belt*, because it transports water, salt, heat, and nutrients throughout the ocean (see Figure 16-10).

FIGURE 16-10:
The global conveyor belt.

It you were able to watch a specific bubble of water during this whole adventure, it would take about 1,000 years for that bubble to complete one full lap.

FUN FACT

The area where the North Atlantic deep water forms (beneath the Denmark Strait, which separates Iceland and Greenland) is actually the world's largest waterfall. As warm water near the surface loses heat to the atmosphere, it sinks, pushing the colder, denser water below it over a ridge about 610 meters (2,000 feet) below the surface. This water then cascades down, plunging to a depth of about 3,050 meters (10,000 feet). However, if you were to ride a submarine over this falls, you would scarcely feel any motion. While the volume of water dwarfs the flow of the Amazon River, the water doesn't flow nearly as fast.

REMEMBER

The global conveyor belt is vital for all life on Earth, and it depends heavily on temperature variations between the poles and the equator. It helps cool the equator and warm areas closer to the poles, making the entire planet more habitable for plants and animals, including us. It also churns up nutrients from the deep ocean, bringing them to the surface and distributing them around the world. If climate change results in significant enough warming (and thus the freshening of seawater) in the polar regions (because no sea ice will form to increase salinity in the water below), these currents would be disrupted, impacting the heat distribution of ocean currents with possibly catastrophic impacts on the rest of the planet. Not good, to put it mildly.

Knowing Where the Winds Blow

Ocean currents and temperatures play a role in driving wind patterns and vice versa. In addition, both currents and winds are influenced by Earth's rotation as well as by geological formations, such as mountains, valleys, and canyons (above or below water).

On a global scale, the sun heats Earth more near the equator and less near the poles, so hot air rises at the equator and travels toward the poles, while cold air sinks at the poles and travels toward the equator. (Remember, hot air rises and cold air sinks.) If this were all that were going on, Earth would have two circulation cells — one for the Northern Hemisphere and one for the Southern Hemisphere.

Unfortunately, that would be too easy due to the uneven distribution of land and water around Earth and due to Earth's rotation, each hemisphere, north and south, has three circulation cells (see Figure 16-11):

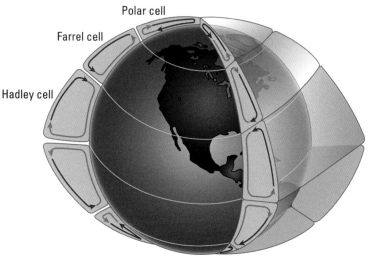

©John Wiley & Sons, Inc.

FIGURE 16-11:
Three
circulation
cells.

>> **Hadley cells:** Near the equator, warm air rises and flows toward the poles, cooling along the way, dropping to the surface, and flowing back toward the equator.

>> **Polar cells:** Cold air over a pole sinks, travels along the surface gathering heat, then rises and flows back toward the pole.

>> **Farrel cells:** Warm air travels toward a Hadley cell and is pulled down toward the surface by the Hadley cell's air flow, cools as it travels along Earth's surface, and is then lifted by the polar cell's air flow. Think of a Farrel cell as a gear between a Hadley and a polar cell.

Where these cells meet create bands of high and low pressure areas — low pressure where air is moving upward and high pressure where it's moving downward.

These circulation cells aren't neatly aligned north and south. They're twisted due to the *Coriolis effect* — the apparent deflection of an object in motion on or near the Earth due to the Earth's rotation. Because Earth is wider at the equator than at the poles, everything at the equator moves faster than at the poles to complete the 24-hour rotation. Now imagine you're standing on the equator and you try to throw a ball straight to your friend who's standing in the Arctic. Because she's moving slower than you (and the ball), the ball will fly straight but land to the east of her as if it had curved and passed her by. If she tries to throw a ball to you, it will land to the west of you, as if it had curved and fallen behind you. The result is that air moving from the equator toward the poles is deflected east, while the air moving from the poles toward the equator is deflected west. (See Figure 16-12.)

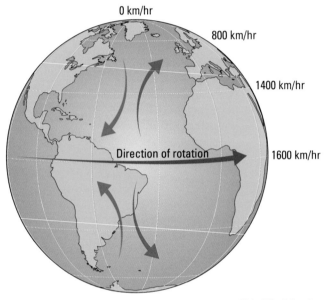

FIGURE 16-12:
The Coriolis effect.

As a result, three bands of prevailing winds form in each hemisphere (see Figure 16-13):

» Trade winds blow east to west near the equator because the air in the Hadley cells flows toward the equator and is deflected west. They're called trade winds because sailors traveling from Europe to the Americas used them to speed up their journey (and still do today).

» Westerlies blow west to east in the middle latitudes because the air in Ferrel cells flows away from the equator and is deflected east.

» Polar easterlies blow east to west in the polar regions because the air in polar cells flows toward the equator and is deflected west.

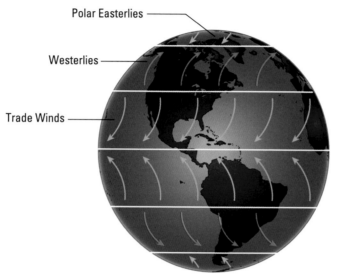

©John Wiley & Sons, Inc.

FIGURE 16-13:
Three prevailing wind bands in each hemisphere.

Going Round and Round with Gyres

All these forces moving the water around impact one another, thereby influencing the direction of the currents. In addition, as currents flow, they eventually bump into landmasses that influence their direction. Within the different ocean divisions (North Atlantic, South Atlantic, North Pacific, South Pacific, and Indian Oceans) currents develop that flow in elliptical patterns called *gyres* — they're sort of like gigantic, slow-moving whirlpools (see Figure 16-14).

These gyres are driven by wind. In the Northern Hemisphere, trade winds blow down and west, while westerlies blow up and east, creating a clockwise flow to wind and water. In the Southern Hemisphere, trade winds blow up and west and westerlies blow down and east, creating a clockwise flow to wind and water. In contrast, the air around hurricanes and cyclones flows toward an intense low-pressure area at the center of the storm, so air flowing from the equator toward a pole is bent toward the east, while air flowing from a pole toward the equator is bent west. This causes hurricanes to spin counterclockwise in the Northern Hemisphere, while cyclones, the Southern Hemisphere's version of hurricanes, spin clockwise.

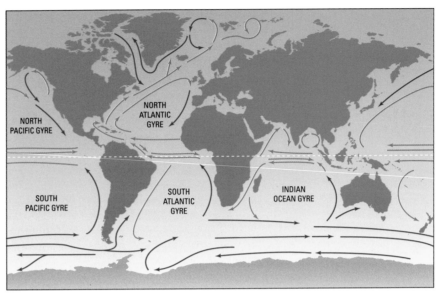

FIGURE 16-14:
Gyres.

©John Wiley & Sons, Inc.

One exception to the formation of gyres is in the Southern Ocean (somebody always has to be different!), where the current flows mostly from west to east around the entire globe. This is possible because the current has no major obstacle, such as a continent, to deflect the water. Another exception is at the equator, where the Coriolis effect is essentially zero and the water flows in the direction of the wind — east to west. These gyres distribute nutrients, impact ocean navigation, facilitate the travel of migratory species, and serve as a place where ocean debris gathers.

FUN FACT

You may have heard that water spirals down a drain clockwise in the Northern Hemisphere and counterclockwise in the Southern Hemisphere. This is untrue. The forces that cause the creation of gyres act on a global scale, but do not affect the water in your sink or toilet.

All five oceans have plastic pollution in them (no corner of the planet has been spared), but the Great Pacific Garbage Patch between California and Hawaii is the largest. Now when we say "patch," we don't mean one huge island of plastic but rather a large region with scattered areas in which high concentrations of plastics and other debris accumulate. In fact, some areas within the bounds of the Great Pacific Garbage Patch may have little or no noticeable debris in the water. And it's not just the pieces of plastic that cause a problem. Over time, plastic breaks up into smaller and smaller pieces, making clean-up exceedingly difficult, but it never goes away, posing a persistent threat to marine life. The best solution is prevention — stopping plastics from entering the ocean in the first place.

Following the Ups and Downs of Sea Levels

Sea level *sounds* stable and easy to determine. Just about everyone on the planet uses it as the basis for describing elevations. Denver is the Mile High City, 5,280 feet above sea level. Mount Everest is 8,850 meters (29,035 feet) above sea level. The Dead Sea is 423 meters (1,388 feet) below sea level. However, as you probably know by now, nothing is that simple; sea level isn't uniform around the entire planet. Many different factors impact sea levels, and these factors change over time, so sea level can vary dramatically.

To understand what sea level really means, first you have to recognize that it differs depending on how it's measured:

>> **Local sea level** is the height of the water along the coast relative to a specific point on land. To measure local sea levels, scientists use tide stations.

>> **Mean sea level** (*tidal datum*) is determined by measuring sea levels every hour in the same spots over a period of 19 years, then averaging all the measurements. Why 19 years? Because that's how long the moon takes to complete a full rotational cycle (the *metonic cycle*), and the moon has the most influence over tides.

>> **Global mean sea level** is the average sea level across all coasts. Sea level isn't uniform around the planet. It is higher in some areas than others. However, global mean sea level is the best attempt at determining an average. To measure changes in global sea levels, scientists use satellite laser altimeters that bounce radio waves off the surface of the ocean at different points and measure the time the radar beam takes to travel back to the satellite.

Monitoring sea level is important, especially for coastal areas, and based on the best information available sea levels are rising. Yep, climate change has raised the global mean sea level by 21 to 24 centimeters (8 to 9 inches) since about 1880, and about one-third of this increase is attributable to the last 25 years (see Figure 16-15). Since 1993, when satellite altimetry was first put into place, sea level has risen on average 3.3 millimeters (about 1/8 inch) each year. In contrast, during the 19th century, sea level rose at an average of less than 1 millimeter (about 1/25 inch) each year.

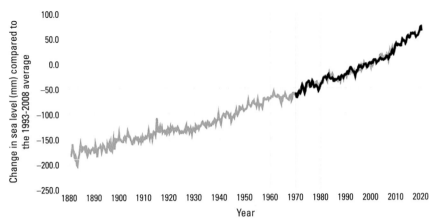

FIGURE 16-15:
Global mean sea level change from 1880 to 2020.

Source: Climate.gov https://www.climate.gov/news-features/understanding-climate/climate-change-global-sea-level

Climate change drives two factors that contribute most to the global rise in sea level:

>> **Meltwater:** As land-based glaciers and ice sheets melt (as they're doing in Greenland and Antarctica, among other places), the extra water flows to the ocean.

>> **Thermal expansion:** When water warms, it expands (physics, baby). Remember, according to NASA the ocean has absorbed 90% of the heat trapped by our emissions of greenhouse gases.

Now a rise in sea level of 3.3 millimeters (1/8 inch) each year may not sound like much. But the rate of sea level rise is accelerating as the planet warms, and many experts fear that sea level will rise almost a meter (3 feet) by the end of this century. The accelerating upward trend is a major concern for several reasons:

>> Sea level affects some areas worse than others, especially small islands that don't have much land to lose (some islands in the Pacific are only a few feet above sea level).

>> A lot of people live on the coasts. Eight of the world's ten most populous cities are on a coast. In the United States, 40 percent of the population lives in coastal cities, and sometimes those cities are only a few feet above sea level already. Huge swaths of New York City, for example, are only 6 feet (about 2 meters) above sea level, and much of Louisiana is less than 3 feet (1 meter) above sea level. As sea level accelerates over the next few decades, high tides and storms will increasingly cause catastrophic flooding, a topic we address in the next chapter.

>> Coastal flooding and storm surges are eroding coastlines, and as sea level rises, that erosion is getting worse and worse, threatening trillions of dollars of real estate.

>> Insurance companies are wise to the risk and aren't so eager to insure coastal homes these days, which is already having a considerable impact on the economies of coastal communities.

The predictions of sea level rise if we don't get a handle on climate change are scary. If you were thinking about getting a house on the beach, we humbly suggest that you reconsider and perhaps opt for a cabin in the woods instead.

A DROP IN SEA LEVEL?

In some places, sea level is actually dropping. How? Well, whenever something heavy is removed from creating downward pressure on the land such as when glaciers melt, that weight loss can cause the land to rise. But this isn't exactly a drop in sea level; it's a rise in land level. The fact of the matter is that global sea level *is rising*, and the problem can be even worse in regions where oil or water is pumped out of the ground causing the land to settle.

Chapter **17**

Driving Climate and Weather

Earth is often thought of as three separate entities — land, air, and water — but certain weather events are regular reminders of how these three entities interact. Hurricanes and cyclones (the hurricanes of the Southern Hemisphere) deliver the starkest reminders, gathering energy and moisture from the ocean into the atmosphere and then unleashing it on any land masses that happen to be in their path.

However, land, air, and water are constantly interacting in less dramatic ways to distribute heat, water, nutrients, and other chemicals around the world, thereby making the planet a habitable place for a diverse collection of life-forms, including us. In this chapter, we focus on how the ocean plays a key role in keeping the planet healthy, specifically with respect to driving climate and weather.

Understanding the Ocean's Role in Climate and Weather

One of the best ways to understand how land, air, and water work together to create climate and weather is to experience it for yourself by spending an entire day and night on the beach. During the day, when the beach is hot and the water is relatively cool, the warm air from the land rises, creating a low pressure area that pulls cooler, denser air from the water over the land. This cool, incoming breeze is called a *sea breeze,* and boy, is it refreshing when you're sitting on a hot beach. At night, the ground loses its heat quickly, while the ocean, which has been absorbing heat all day from the sun, releases it much more slowly into the atmosphere. This causes a low pressure area above the water, which pulls cooler air from the land out away from the beach, creating a *land breeze.*

These beach breezes are just one example of how heat energy moves between land, air, and water. The ocean is in a continuous process of absorbing energy from the sun, storing it in the form of heat, distributing that heat throughout the oceans via currents (see Chapter 16), and exchanging heat with the atmosphere above it. As the ocean absorbs and releases energy, and as air rises and falls and blows around the globe, these forces interact to establish climates and create weather patterns.

Differentiating climate and weather

Any discussion of climate and weather needs to begin by defining those terms:

» **Climate:** The prevailing weather over a long period of time (usually more than 30 years), such as *tropical* (warm and wet), *desert* (hot dry), *polar* (cold and dry), and *temperate* (neither extremely hot nor extremely cold).

» **Weather:** Atmospheric conditions over a short period of time with respect to temperature, sunshine, storms, wind, and *precipitation* (rain, snow, sleet, hail). One day may be sunny and warm and then the next cool, rainy, and overcast.

REMEMBER

NASA defines the difference between climate and weather perfectly — "Climate is what you expect, like a very hot summer, and weather is what you get, like a hot day with pop-up thunderstorms."

Essentially, climate is weather averaged over several decades. However, areas can also experience short-term changes to climate driven by certain events, such as volcanic eruptions or temporary changes in ocean currents, such as El Niño and La Niña (see the later section "El Niño and La Niña").

Looking at how the ocean impacts climate and weather

The ocean plays a huge role in keeping Earth warm by capturing, storing, and distributing the sun's energy. All day every day, some portion of the ocean is getting blasted by the sun's rays, and like a solar panel, it captures this radiant energy, which it then stores in the form of heat. Think of it as putting a pot of water on a burner to heat; the energy is transferred from the burner to the water just as energy is transferred from the sun to Earth.

The part of the ocean where the sun's rays at midday strike Earth from nearly overhead (along the equator) captures and stores the most energy, whereas the polar regions (where the sun never gets very high in the sky) capture the least. The ocean's currents (explained in Chapter 16) circulate the water, so cold water that upwells from the deep ocean and the export of heat from the tropics keeps some regions from overheating, while warm water from the tropics helps to warm the rest of the planet. Evaporation also plays a key role in distributing heat, cooling warm surface waters while increasing the temperature and humidity of the surrounding air, which in turn forms clouds and storms, which return that moisture to the surface.

Letting Off Some Steam

As you heat a pot of water, the energy added to the water makes it evaporate faster and faster, producing more and more steam. You can look at steam as a way that water releases excess energy. Ocean temperatures aren't even close to reaching the boiling point, but the ocean is constantly "letting off steam" through circulation and evaporation. In this section, we look at a few ways the ocean lets off steam that are dramatic enough to catch the attention of meteorologists.

Hurricanes, cyclones, and typhoons

The most dramatic way the ocean lets off steam is in the form of tropical storms — hurricanes, cyclones, and typhoons. Technically referred to as *tropical cyclones*, they're basically all the same — massive spiraling storm systems that form, travel, and intensify over warm tropical ocean waters (typically about 27 degrees Celsius, which is about 80 degrees Fahrenheit). They can reach 16 kilometers (10 miles) high and be more than 1,600 kilometers (1,000 miles) in diameter. Tropical cyclones are distinguished mostly by where they occur. In the North Atlantic and Northeast Pacific Oceans and in the Caribbean Sea, they're called *hurricanes*; in and around the Indian Ocean and Southwest Pacific, they're called *cyclones*; and in the Northwest Pacific, they're called *typhoons*. Another difference is that these storm systems spin counterclockwise north of the equator and clockwise south of the equator due to the Coriolis effect (see Chapter 16).

These tropical storms are all driven by ocean energy and the dynamic between hot and cold. Remember, hot, moist air is less dense than cold, dry air, so hot, moist air creates low pressure and cold, dry air creates high pressure. Because nature abhors a vacuum, wherever there is low pressure (hot, moist air), more air is drawn in to equalize the pressure. Over the ocean, that warm, moist air evaporates from the surface and rises, creating an area of low pressure beneath it as it moves upward. Surrounding air at the surface of the ocean spirals in toward the area of low pressure, where it then warms and rises, creating a spinning pillar of air. The core of this pillar (the *eye*) is very low-pressure air that's calm and clear. As more and more energy (in the form of heat) is added from the warm surface waters below, the pillar grows in size and intensity (see Figure 17-1).

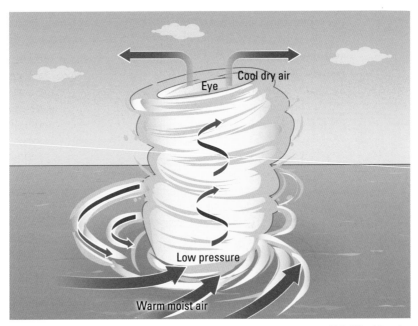

FIGURE 17-1: Formation of a tropical cyclone.

©John Wiley & Sons, Inc.

Every tropical cyclone begins over warm water as a *tropical disturbance* — a tropical weather system of organized *convection* (circulation of air, with warmer, lighter air rising and colder, denser air sinking). This tropical disturbance sometimes develops into a *tropical depression,* defined as rotating thunderstorms with winds of 61 kmph (38 mph) or less. If it continues to develop, the tropical depression turns into a *tropical storm* when the winds hit 63 kmph (39 mph). And a tropical storm becomes a *tropical cyclone* (the formal term for a hurricane, cyclone, or typhoon) when its winds reach 119 kmph (74 mph).

Tropical cyclones are ranked by strength according to the Saffir-Simpson Hurricane Scale:

- » **Category 1:** Winds 119–153 kmph (74–95 mph) — faster than a cheetah

- » **Category 2:** Winds 154–177 kmph (96–110 mph) — as fast or faster than a baseball pitcher's fastball

- » **Category 3:** Winds 178–208 kmph (111–129 mph) — similar, or close, to the serving speed of many professional tennis players

- » **Category 4:** Winds 209–251 kmph (130–156 mph) — faster than the world's fastest rollercoaster

- » **Category 5:** Winds more than 252 kmph (157 mph) — close, to the speed of some high-speed trains

FUN FACT

The World Meteorological Organization maintains a rotating list of names for tropical cyclones that are appropriate for each tropical cyclone basin. Names are assigned to distinguish between cyclones occurring at the same time and to provide a convenient way to reference them in warning messages. Each year, tropical cyclones receive names in alphabetical order, alternating the use of men and women names. When a tropical cyclone causes a lot of damage, its name is retired from the list, so thankfully no more Katrinas, Harveys, or Sandys.

INCREASING DEVASTATION

From 1980 to 1989 (nine years) the U.S. experienced six hurricane disasters resulting in a total of $38.2 billion in damage. In contrast, from 2016 to 2019 (three years), the U.S. experienced seven hurricane disasters resulting in a whopping $335 billion in damage. (These numbers are from the National Oceanic and Atmospheric Administration, and the dollar values have been adjusted to account for changes in the cost of living.)

This massive increase in damage over the past few years is due to multiple factors, but here are a few of the ones with the greatest impact:

- Increased coastal development, which places buildings in the path of tropical cyclones and storm surges

- Destruction of natural barriers against storm surge, such as barrier islands, reefs, seagrass meadows, estuaries, mangroves, and more

- Climate change, resulting in ocean warming, which increases the frequency and strength of tropical cyclones (see the later section "Understanding Climate Change and Global Warming").

Though tropical cyclones can be extremely powerful storms on the water (you don't want to be in a boat that's in the path of one of these monster storms), they don't really bother marine life below the surface. It's when they make landfall that all the damage you see in the news occurs. Not only do these tropical cyclones pound land with torrential rain and powerful wind for hours or even days on end, but they also create *storm surge*, pushing a massive amount of water ashore, causing extreme flooding.

Monsoons

A *monsoon* is a season of prevailing wind that brings either rain or dry air to a tropical region over the course of three to four months. Monsoons are caused by a difference in temperature and air pressure between the ocean and adjacent land.

In the summer, when the land is warm, low pressure develops over land that draws warm, moist air from the ocean, causing a rainy season, which is called a *wet monsoon* (or summer monsoon). Depending on the area, it can produce hundreds of inches of rain in a few months' time. The record for one season was 2,659 centimeters (1,047 inches, which is over 87 feet!) in Cherrapunji, India, in 1860. The *dry monsoon* (winter monsoon) occurs when the land cools off faster than the water, resulting in high pressure over the land that pushes the damp ocean air away, resulting in a period of dry weather.

Monsoons occur in the *Intertropical Convergence Zone* (ITCZ) — a narrow band near the equator where northern and southern air masses converge. The ITCZ shifts north and south in a seasonal cycle (see Figure 17-2). This shift generally brings more rain to northern areas of the tropics from June to September and to southern areas of the tropics from December to February. However, rainfall varies locally due to other factors, such as currents and wind patterns.

REMEMBER

Wet monsoons are vital for regional populations, restoring their water reservoirs, watering their crops, and keeping their hydro-electric power plants running. Unfortunately, as climate and ocean currents change, so does the reliability of rainfall. Too much rain leads to massive flooding, crop failure, deadly landsides, and the drowning of people and livestock. Too little rain leads to droughts, crop failure, livestock starving, and loss of income and electricity.

El Niño and La Niña

The *El Niño-Southern Oscillation* (ENSO) is a largescale ocean-atmosphere climate interaction linked to a periodic warming in sea surface temperatures across the central and east-central equatorial Pacific. This unseasonably warm ocean water was first recognized by fishermen off the coast of South America in the Pacific

Ocean during the 1600s. The phenomenon was named El Niño (Spanish for "little boy") because it was around Christmastime and was named after the Christ Child. But honestly, every time *we* hear "El Niño," we think about Chris Farley playing it on *Saturday Night Live.*

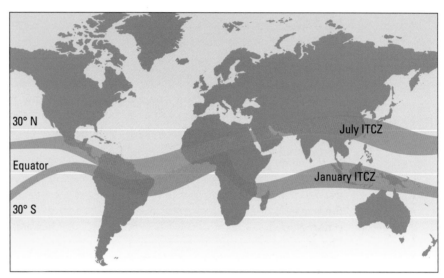

FIGURE 17-2: Seasonal shift of the ITCZ.

During normal non-El Niño years, trade winds blow from east to west, across the middle of the Pacific Ocean along the Equator from the Americas towards Asia. The winds push warm water west and it piles up against Asia so much so that the water level off Indonesia is about one and a half feet higher and about 8 degrees warmer than it is off the coast of Ecuador. This produces an upwelling of cold, nutrient-rich water that flows to the coast of South America (which supports super productive fishing in that area). The trade winds carry water vapor to Asia and leave South America dry and cool.

But every two to seven years this pattern reverses (see Figure 17-3). As the trade winds relax, warm water flows eastward, toward South America. Sea levels fall on the west side of the Pacific, and they rise on the Pacific coast of the Americas. Places like the Galapagos Islands, Peru, and Southern California are visited by much warmer water with disastrous effects. During the El Niño of 1982–1983, the Galapagos lost about 96 percent of all their corals. In 1998, El Niño caused thousands of sea lions in California and Peru to perish, as well as countless dolphins and whales. In addition to the warm water, rain is carried and dumped on the coast of the Americas (which is usually dry), resulting in an increased risk of devastating floods and landslides. At the same time, in the Indo-Pacific, the lack of rain causes serious droughts and forest fires and weakens the regular monsoons that Asia and India rely on.

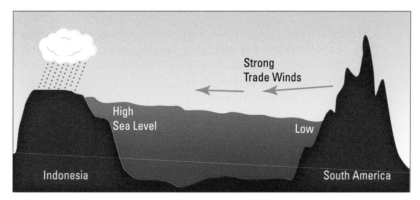

Normal Conditions

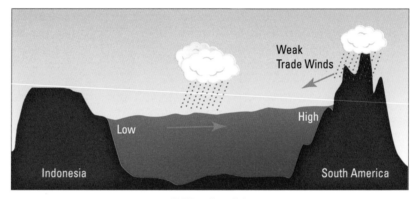

FIGURE 17-3:
El Niño
reverses
course.

El Niño Conditions

La Niña ("the little girl") is the opposite of El Niño. It's characterized by unusually low ocean temperatures in the equatorial Pacific. Though it's not as widely talked about, La Niña can do a lot of damage too, resulting in cooler, wetter air flowing over the northern part of the U.S. (damaging crops) and warmer, drier air across the southern portions of the U.S. and northern Mexico (causing drought and increased frequency and severity of fires).

Understanding Climate Change and Global Warming

Nature has done a remarkable job of maintaining conditions suitable for life on this planet to evolve and flourish. It's done this, to a large degree, by maintaining a stable carbon cycle and a healthy balance between the carbon in the ground, air, and water and in plants and animals.

Unfortunately, human activities have disrupted this cycle by extracting carbon-based, "fossil" fuels (coal, oil, and natural gas) from the ground and burning them, pumping massive amounts of carbon dioxide into the atmosphere that would otherwise be sequestered in the ground. At the same time, people have been destroying large ecosystems such as forests, mangroves, and seagrasses, which also absorb and store carbon. When forests and other ecosystems are destroyed, much of the carbon contained in their biomass ends up as even more carbon dioxide in the atmosphere.

The carbon dioxide produced by the burning of fossil fuels and forest destruction, along with *greenhouse gasses* produced by agriculture (such as methane and nitrous oxide) as well as other greenhouse gasses (like chlorofluorocarbons that can leak from old refrigerators and air-conditioners where it's known as Freon), act as a blanket that prevents heat from escaping Earth's atmosphere. The result is *global warming* — the long-term heating of Earth's climate due to human activities.

Global warming is largely driven by the *greenhouse effect* (see Figure 17-4). Think about what happens when you leave your car in the sun with all the windows rolled up. The sunlight passes through the windows and is absorbed by everything inside the car, but because the windows are rolled up, the heat can't escape, so heat builds up inside. With global warming, sunlight passes through the atmosphere and, is absorbed by Earth (mostly the ocean). Normally some of that heat would escape back into space. But, due to more and more greenhouse gas molecules absorbing more and more of that heat our planet continues to warm.

Global warming is responsible for *climate change,* the complex shifts in our Earth's climatic systems, which cause a myriad of impacts such as rising sea levels, accelerating ice melt in polar regions, changes in long-term weather patterns, and other measures of climate occurring over several decades or longer due to the excess heat trapped in our atmosphere.

NOT SO NEW

The first calculation of the greenhouse effect to include human-driven release of greenhouse gases happened over 100 years ago. Swedish chemist Svante Arrhenius estimated that the doubling of CO_2 content in the planet's atmosphere would raise its temperature by 2.5 to 4.0 degrees Celsius (4.5 to 7.2 degrees Fahrenheit). So the science isn't new and neither is the fossil fuel industry's funding of pseudo-science in denial of it. The same scientific method used to create modern medicine, put humans into space, and generally create the world we take for granted is what has been used to predict climate change for over 100 years. The time for denial is long since over; now it's time to act.

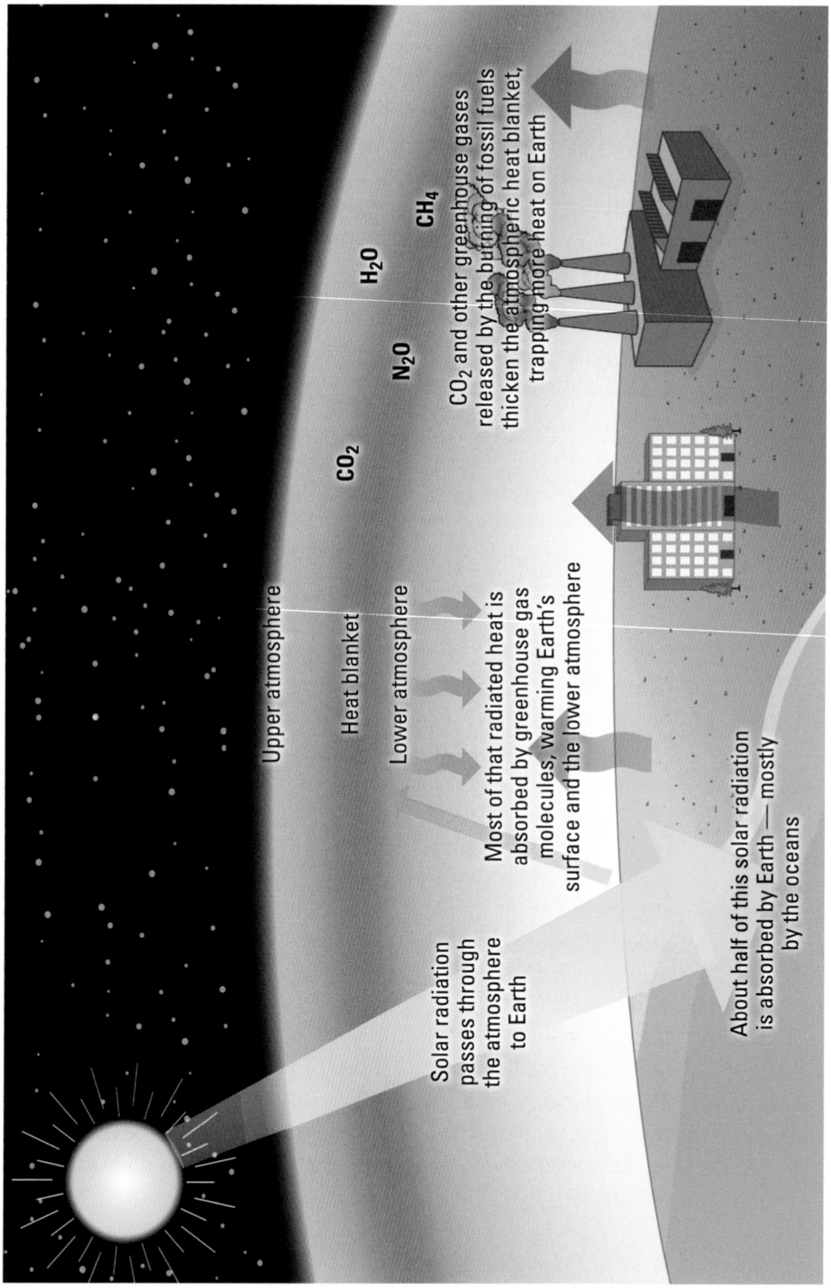

© John Wiley & Sons, Inc.

Upper atmosphere

Heat blanket

Lower atmosphere

CO$_2$

N$_2$O

H$_2$O

CH$_4$

CO$_2$ and other greenhouse gases released by the burning of fossil fuels thicken the atmospheric heat blanket, trapping more heat on Earth

Most of that radiated heat is absorbed by greenhouse gas molecules, warming Earth's surface and the lower atmosphere

Solar radiation passes through the atmosphere to Earth

About half of this solar radiation is absorbed by Earth — mostly by the oceans

FIGURE 17-4: The mechanics of global warming.

The ocean has been absorbing much of this excess heat — about 90 percent by some estimates — which has served to slow down the rate of global warming and climate change. However, the symptoms of global warming (climate change) are becoming more and more obvious the higher the ocean's temperature rises:

>> Flooding, droughts, and tropical storms are becoming more frequent and more severe, leading to starvation, conflict over dwindling natural resources, and mass migrations of people.

>> Snowpack (which acts like a natural reservoir slowly feeding water from mountain ranges and polar areas into agricultural areas and cities throughout the summer) is reduced, and the snow melts faster.

>> Glaciers and sea ice in polar regions are melting faster.

>> Permafrost is melting, which has the potential to release a lot of methane (another powerful greenhouse gas) into the atmosphere.

>> Ice sheets in Greenland and Antarctica are melting seven times more rapidly than just two decades ago, increasing sea levels.

>> Coral reefs are dying at an unprecedented rate with 40 percent of the world's coral reefs already lost.

>> The hydrogen ion concentration in surface seawater has increased by about 50 percent. In other words, the seawater has become less alkaline because it can't keep up with the amount of carbon dioxide in the atmosphere, negatively impacting marine species that need more alkaline seawater to build shells to survive. (This is also known as ocean acidification.)

>> Species distribution is being altered by changes in the environments in which they live.

REMEMBER

The bottom line is that our ocean is in big trouble, meaning that life as we know it is in big trouble. Our very existence is intricately tied to the fate of the ocean. Trust us, if you like breathing, eating, mild sunny days with a cool breeze, and an awesome planet teeming with beautiful and fascinating wildlife, then the health of the ocean matters to you. Simply put, as stewards of this planet, we must take care of the system that takes care of us. For ways to do just that, see Chapters 19, 21, and 24.

IT'S OFFICIAL

The past five years (2014–2019) have been the warmest years ever recorded for the ocean, with 2019 being the warmest yet. A gigantic study published in 2019 by the United Nations Intergovernmental Panel on Climate Change, entitled the *Special Report on the Ocean and Cryosphere in a Changing Climate*, looked at how the ocean and *cryosphere* (the parts of Earth where water is frozen) have been and are expected to change with ongoing global warming, the risks and opportunities these changes bring to ecosystems and people, and mitigation, adaptation, and governance options for reducing future risks. According to the IPCC report:

"It is virtually certain that the global ocean has warmed unabated since 1970 and has taken up more than 90% of the excess heat in the climate system (high confidence). Since 1993, the rate of ocean warming has more than doubled (likely). Marine heatwaves have very likely doubled in frequency since 1982 and are increasing in intensity (very high confidence). By absorbing more CO_2, the ocean has undergone increasing surface acidification (virtually certain). A loss of oxygen has occurred from the surface to 1,000 m (medium confidence)."

So don't listen to the doubters or the conspiracy theorists. Global warming and climate change are real, they're happening right now, and they threaten us all. For the full report, visit www.ipcc.ch/srocc.

5

Understanding the Human-Ocean Connection

Chapter **18**

Taking a Quick Dip into the History of Underwater Exploration

"Space: The final frontier . . . ," right? Not so fast, Captain Kirk. Before we launch our "five-year mission to explore strange new worlds," let's get back down to Earth. We still have loads of it to explore, most of which is below the surface of the sea. And if the goal is to "seek out new life," you need go no farther than about seven miles beneath the surface of the ocean. By one estimate, about 8.7 million species are waiting to be discovered on *this* planet, many of which live in the ocean.

Fortunately, ocean explorers and inventors have been hard at work cracking the mysteries of the ocean and increasing our understanding of it, and many of the advances can be credited to technologies that have been developed over the past 200 years or so. Sure, ocean exploration was going on prior to the 1800s, but most of it was done just to find more land. The ocean was seen as an expanse to be traversed between land masses rather than a wondrous world of its own.

We're not slighting the ocean explorers and seafarers of old. What they accomplished with limited tools and technologies was amazing, but they merely skimmed the surface. If they could have plunged the ocean depths, we're sure they would have, but they faced two insurmountable obstacles — an inability to breathe underwater and intense water pressure. Since then, humans have developed all sorts of technologies for exploring the ocean, and in many ways, the history of ocean exploration can be told through the development of these technologies.

Getting to the Bottom of Things

In the early 1800s, most undersea exploration was performed from the surface via ocean *surveys* (studies) and *soundings* (depth measurements) from ocean-going vessels, and they were conducted mostly to understand the physical characteristics of the ocean, including tides, currents, ocean depths, and elements of the seafloor. Of course, even long before the 1800s, people were fishing, swimming, and free-diving in the ocean, so they had a rudimentary understanding of what marine life was like — at least as far down as about 50 meters (150 feet).

However, a fuller understanding of the ocean and marine life requires deeper dives and observing creatures living and interacting in their natural habitats (what you pull up from the deep is dead once it hits the surface). You can't do that if you can perform only shallow dives and stay underwater for only a few minutes at a time. Advances in ocean exploration required innovations that would overcome these limitations.

Diving bells

To go underwater, you need to hold your breath, right? Maybe not. What if you could bring a little breathing room with you? That's the idea behind the diving bell, an invention written about by Aristotle back in the fourth century B.C.E. Imagine turning a cup upside down and pushing it straight down into a pool full of water. As long as you don't let it tip, the water pressure keeps the air from escaping the cup. Now imagine a very large, heavy cup big enough for you to stand or sit inside — that's a diving bell. The bell is lowered into the ocean, with you inside it, breathing comfortably as it's lowered deeper and deeper. You can look down to observe what's in the water below you or take a deep breath, swim outside the bell to explore, and then swim back when you need more air. Legend has it that Alexander the Great explored the Mediterranean in a clear glass diving bell.

Unfortunately, a diving bell is one of those inventions that looks great on paper but isn't quite ready for prime time. In fact, diving bells can even be dangerous for several reasons:

>> They tend to tip, in which case they lose the air required to keep the diver alive.

>> Without fresh air being pumped into the bell from above, the bell fills with carbon dioxide from the air the diver exhales, which can become toxic.

>> People back then didn't understand the risks of surfacing too quickly after a dive. (See the nearby sidebar on decompression sickness.)

Diving bells are still in use today, usually to transport divers back and forth between the deck of the boat and wherever they need to work below the surface. However, these bells and how they're used have been modified. They now have ballasts to keep them from tipping and their own supply of compressed air (or hoses to circulate air from the surface). When returning divers to the surface, they're pulled up slowly to prevent decompression sickness.

UNDERSTANDING DECOMPRESSION SICKNESS

Decompression sickness, also called *generalized barotrauma* or "the bends," occurs as a result of a rapid decrease in the pressure around you, which happens when you ascend from a dive or go up in an airplane. However, most aircraft have pressurized cabins to reduce the effect. When you scuba dive and are breathing compressed air, your body takes in extra oxygen and nitrogen. Your body uses the oxygen (yay), but your blood absorbs the extra nitrogen (boo). As you swim back toward the surface, the water pressure decreases. If this decrease in pressure happens too quickly, the nitrogen doesn't have time to clear from your blood slowly and instead comes out too quickly and forms bubbles in your blood and tissues.

While bubbles in the bathtub are great, nitrogen bubbles in your body are not. They can cause a whole lot of adverse reactions, such as pain, headaches, rashes, and even death. So, the longer and deeper you dive, the more gradually you need to ascend to give the nitrogen sufficient time to slowly leave your body.

Fun fact: Decompression sickness is sometimes called "the bends" because in serious cases it can cause physical trauma in your joints that makes you *bend* over in pain.

Hard-hat diving helmets and suits

Diving bells work, but they're expensive and cumbersome, as they were when they were first put to use. So, what if we shrunk the bell down and put it on the diver's head? Welcome to the hard helmet and hard-hat diving suit.

The first successful diving helmets were made by Charles and John Deane in the 1820s. The original design was for a smoke/fire helmet to help firefighters breathe while battling flames. Later, the Dean brothers tweaked it to supply air to divers underwater. It was a hard helmet that had a single hose attached through which air was pumped from the surface down to the diver. The bottom of the helmet wasn't attached to anything, allowing air to escape along the bottom of the helmet. Unfortunately, if the diver bent over (or fell over), the helmet quickly filled with water.

Later, a German inventor named Augustus Siebe refined the design by developing a way to seal the helmet to an airtight diving suit and improving the exhaust system, which led to the "Siebe Improved Diving Dress." Unfortunately, early helmets weren't equipped with non-return valves, so when a hose to the helmet ruptured or was severed, the escaping air caused extreme and sudden negative pressure resulting in sometimes deadly "diver squeeze." In some cases, the negative pressure was extreme enough to suck the skin off the diver's head and into the helmet. Seriously, it happened, more than once . . . not a good way to go.

Those hurdles were overcome (no more skin sucking), and hard helmet diving was officially adopted by the Royal Navy in the 1840s. Subsequent variations of the Siebe Improved Diving Dress were developed with the U.S. Navy (see Figure 18-1), and various designs of the "copper dive helmet" pioneered more than 100 years earlier was used well into the 1970s.

Modern hard helmet designs continue to evolve and are still used today for commercial diving, underwater welding, hazardous diving, and other applications (see Figure 18-2).

The self-contained underwater breathing apparatus (SCUBA)

While Jacques Cousteau is Philippe's grandfather, Ashlan (who wrote this chapter) is only going to reference him in a historical context here.

A few pivotal years in the 20th century signaled a new beginning for humankind: 1903, for example, when the Wright brothers flew their first airplane and 1961 when Yuri Gagarin became the first person in space. In the history of undersea exploration, 1943 was such a year — the year when a young French Naval Officer named Jacques-Yves Cousteau together with an engineer named Emile Gagnan, tested and patented a new type of underwater breathing apparatus.

FIGURE 18-1:
Diver Terry
Rioux in a
copper Navy
Mark V dive
suit circa
1970s. The
Mark V design
dates from
around 1916.

Source: Woods Hole Oceanographic Institution – www.whoi.edu

FIGURE 18-2:
Philippe Jr. in a
modern hard
helmet
"hazmat"
diving suit
diving into
an oil slick
during the BP
Deepwater
Horizon oil
spill 2010.

Source: Philippe Cousteau

Unlike its predecessor, the hard helmet diving suit, which tethered the diver to a ship at the surface, this new apparatus was an *autonomous* (untethered) system that featured an on-demand regulator attached to a tank of pressurized air. Cousteau and Gagnan called their invention the *scaphandre autnomne* (French for "autonomous diving suit") and later renamed it the Aqua-Lung (see Figure 18-3).

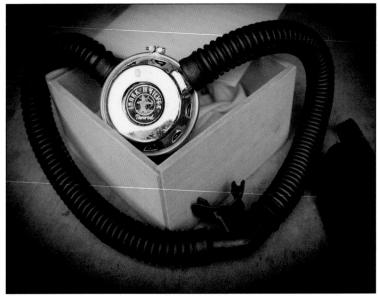

Source: Philippe Cousteau

FIGURE 18-3:
A vintage double-hose regulator.

For the first time, humans were able to swim freely in the ocean for more than a few minutes at a time, and a new era of undersea exploration was created — what we call scuba diving today. Before this innovation, people had to either hold their breath or don a heavy metal helmet and boots and trudge along the bottom of the seafloor. Prior to the widely available Aqua-Lung, most people had no idea what lurked beneath the surface of the ocean's vast expanse. Most of what we knew about the ocean's inhabitants was what we pulled out in fishing nets and scientific trawls.

By 1945, the first commercial regulator, the CG-45, went into production, and scuba diving became a recreational sport. But it didn't stop there — the key to scuba diving's popularity was the proliferation of TV shows and movies that came out in the subsequent years. Jacques Cousteau's first major film, *The Silent World*, which aired in 1956, revolutionized humanity's understanding of the ocean and thrilled audiences in theaters around the world, winning a Palm D'Or at the Cannes Film Festival and an Academy Award in Hollywood, bringing the secrets and splendor of the ocean and its inhabitants to audiences around the world.

Later, his eponymous series, *The Undersea World of Jacques Cousteau*, aired for decades in over a hundred countries around the world. That was the genius of Jacques Cousteau; not only did he co-invent scuba diving, but his films and books also popularized the ocean for generations and thrust the idea of ocean exploration and conservation into the mainstream of public consciousness. Later, he was joined by his sons, Philippe Cousteau Sr. and Jean-Michel Cousteau.

By the late 1960's Philippe Cousteau Sr. had joined the family business and began to pioneer the idea of ocean conservation not just exploration. Philippe Sr. filmed and directed 26 episodes of the *Undersea World of Jacques Cousteau*, his own series *Oasis in Space*, co-authored *Sharks* with his father, toured the world lecturing and more. Philippe was often joined by his wife, Jan, an unsung hero of his expeditions (see Figure 18-4). Through his work, Philippe Sr. had an enormous impact on ocean exploration and conservation even though his life and work were cut short.

FIGURE 18-4:
Philippe Sr.'s crew on expedition in the arctic. Jan (center bottom) with Walrus pup and Philippe Sr. (third from left).

Source: Jacques Renoir

Tragically, Philippe Cousteau, Sr., was killed in an airplane accident in 1979 while on expedition, leaving Jan and their 3-year-old daughter, Alexandra, and unborn son, Philippe Jr., behind. At the time of his death, according to a *Time* magazine poll, Philippe Sr. was the sixth most famous person in the world.

While diving has evolved since those early days, the fundamental mechanics of the modern regulator are more or less the same as the early devices from the 1940s.

Submersibles

Submersibles are small watercraft designed to operate underwater, typically for the purpose of research and exploration. This group of watercraft contains a wide variety of vessels, but, perhaps surprisingly, *not* submarines. Why not? Because *submarines* (subs for short) are designed for independent navigation underwater for long periods of time and are generally naval vessels used for warfare and equipped with torpedoes and/or missiles, and sometimes Sean Connery. Submersibles, on the other hand, are usually tethered to or launched and controlled from (and dependent on) a mothership. The main difference is that submersibles are *not* autonomous, which doesn't mean they're any less cool or fun to ride around in. In this section, we highlight some of the most notable submersibles of the group.

The Bathysphere

This metal sphere was built by engineer Otis Barton and naturalist William Beebe between 1928 and 1929. Weighing in at 4,500 pounds, this hollow steel ball had three small windows, each made of 3-inch-thick quartz, miles of heavy cord that kept the ball tethered to its mothership, a 400-pound exit hatch that had to be sealed shut from the outside, and about eight hours of air (see Figure 18-5). Talk about a potential ball of death!

HANS HASS

The explorer Hans Hass pioneered underwater photography in the 1930s and, in 1939, produced what is widely regarded as the first underwater documentary, *Stalking Beneath the Sea*. In the early 1940s, he worked with an engineer to develop a rebreather diving device. His work is not as widely known, but his contributions to the world of diving are profound nonetheless.

FIGURE 18-5:
A full-size
replica of the
Bathysphere at
Mystic Seaport
Aquarium.

Source: NOAA, Public Domain

After only two unmanned tests (the first of which didn't go so well), Beebe and Barton hopped into their Bathysphere and performed their first deep dive, reaching a depth of 245 meters (803 feet). They continued to dive multiple times in many locations, documenting the sea life they saw along the way. At first, people thought the animals they were describing were delusions, but finally people started to believe the divers' reports. Beebe and Barton continued making history, and on April 15, 1932, they set a record with their deepest dive to 923 meters (3,028 feet).

Bathyscaphe Trieste

Built by Swiss inventor, physicist, and explorer Auguste Piccard, the Bathyscaphe was modelled after the balloon he invented and flew to the stratosphere in 1938. Allegedly, after befriending Beebe, Piccard modified his invention to go down rather than up. And unlike the Bathysphere, which was tethered to a boat above, the Bathyscaphe was a free-diving, self-propelled deep-sea submersible (yes, more like a submarine).

After the original Bathyscaphe was destroyed by a storm, a second was built named the Trieste. It consisted of an area for people that was mounted below a huge balloon-like float filled with gasoline for buoyancy control. After many successful dives in the Mediterranean, the Trieste was bought by the U.S. Navy. On January 23, 1960, the Trieste, carrying Piccard's son, Jacques Piccard, and U.S. Navy Lieutenant Don Walsh, set a new world record by diving to the deepest part of the Ocean, the Challenger Deep at 10,912 meters (35,800 feet) in the Mariana Trench near Guam. The descent took 4 hours and 47 minutes and allowed the two explorers to spend about 20 minutes on the ocean floor. The return trip took 3 hours and 15 minutes. It was an absolutely incredible feat that has been repeated only a few times in history.

Alvin

Talk about a workhorse! This piloted submersible (also known as a Human Occupied Vehicle or HOV) has been in service since 1964 and is still going strong. Able to hold three people (one pilot and two passengers), Alvin has banked countless dives (okay, we're sure someone is counting) all over the world and has truly changed what we know about our ocean and the species that call it home. Operated by Woods Hole Oceanographic Institution, this cool vessel can hover in the water, maneuver over rugged topography, or rest just above the seafloor thanks to seven reversible thrusters — that's quite a feat for a vehicle weighing 45,000 pounds (see Figure 18-6). Alvin can even dive for up to 30 days in a row before requiring a scheduled maintenance. Yes, she's a BEAST.

Remote-controlled submersibles

While piloted submersibles are all the rage, sometimes the job calls for a remote-controlled vessel — an underwater drone (see Figure 18-7). They can be tiny or big and general-purpose vehicles or specialized for taking water samples, collecting species, sampling the ground, testing pH/salinity/turbidity, checking out a wreck site, and more. And without humans inside, they can save on space, don't need air, and needn't be pressurized. These remote-controlled ocean explorers come in two models:

>> **Remotely Operated Vehicles (ROVs)** are machines tethered to a ship or platform at the surface by cables, which allow an operator to control the vehicle remotely.

Source: Woods Hole Oceanographic Institution –www.whoi.edu

Source: Mountains in the Sea Research Team; the IFE Crew; and NOAA/OAR/OER Licensed under CC BY 2.0

>> **Autonomous Underwater Vehicles (AUVs)** are preprogrammed to perform a specific task. They're dropped off from a mothership and return to a preprogrammed location for pickup.

FREEDIVING: BEFORE ALL THE FANCY TECHNOLOGY

Freediving is diving under water without the use of a breathing apparatus. While this may not sound super interesting or even pleasant to average folks, who can hold their breath for only about 30 seconds, it can be an amazing way to explore the ocean. Freediving has been around for as long as people have had a craving for seafood and pearls.

Take the Japanese Ama divers, for example. They've been deep-diving in the ocean in search of mollusks, abalone, shrimp, and pearls since about 925 CE (around 1,100 years). The Ama are in the sea for an average of four hours daily, making about 100 to 150 dives a day, just holding their breath. As you can imagine, they're incredible athletes, and almost all these divers are women.

The Bajau people of Southeast Asia are also accomplished divers. Spending most of their lives living on boats, coming ashore only to sell their catch, they freedive for about five hours a day to depths of more than 70 meters (230 feet) — no wetsuits, no air tanks, no flippers, just homemade goggles and some epic lung control. The Bajau (also called Sea Nomads) have been living this ocean life for the past 1,000 years. Over that time, they've developed spleens that are 50 percent larger than those of the rest of us, which gives them the capacity to carry more oxygen than the rest of us. (When you hold your breath, your spleen releases red blood cells, which carry oxygen to other parts of your body. Deep-diving mammals benefit from having enlarged spleens.)

By the way, the Guinness World Record for breath holding was set in 2016 by Aleix Segura Venrell (a professional freediver) in a shallow pool in Spain. He held his breath for 24 minutes and 3 seconds! He achieved this incredible feat by breathing oxygen-rich air prior to going underwater and then putting his body and mind into a Zen-like state before dipping below the surface. We salute you, sir.

Setting Up Shop in Underwater Research Stations

Decades before the launch of the International Space Station (ISS), ocean explorers were setting up underwater habitats and research workstations in the ocean. Here, we introduce you to some of the most successful programs.

Conshelf

In the 1960s, Jacques Cousteau and his team constructed three Continental Shelf Stations, known as Conshelf, to investigate the effect of living underwater on human physiology:

>> **Conshelf I** was constructed in 1962 in 10 meters (33 feet) of water off the coast of Marseilles, France. The habitat, also known as Diogenes, after the Greek philosopher, was a watertight steel cylinder 5 meters (16 feet) long and 2.5 meters (8 feet) in diameter. The following year, oceanauts Albert Falco and Claude Wesley became the first to spend a week beneath the surface of the sea, during which time they made basic observations of the marine life surrounding the underwater habitat.

>> **Conshelf II,** located in the Red Sea off the coast of Sudan, was considerably more advanced than its predecessor. In addition to the starfish-shaped living habitat submerged at a depth of 10 meters (33 feet), Conshelf II included a deeper habitat at 33 meters (100 feet) and a submarine hangar, which held the team's "diving saucer," a two-person submarine. This time, five oceanauts took up residence in the main habitat for 30 days, during which time two of them spent a week in the deeper habitat. Experiments conducted during the Conshelf II mission made considerable inroads into understanding the physiological effects of diving and pressure and were recorded in the Academy Award-winning 1964 documentary, *World Without Sun.*

FUN FACT

One of my favorite discoveries during this experiment (among so many) was that champagne doesn't bubble or fizz at this depth because of the pressure.

>> **Conshelf III,** led by Philippe Cousteau, Sr., was the final and perhaps most ambitious of all. Six oceanauts spent three weeks in a facility submerged at a depth of over 100 meters (330 feet) in the Mediterranean Sea, between Nice and Monaco. That's quite a long time at a depth of 100 meters.

SeaLabs

Following Conshelf, the U.S. Navy launched SeaLab — an experimental program developed to further test the effects of saturation diving on humans. (*Saturation* involves staying underwater long enough to bring all the tissues of the body into equilibrium with the gasses being breathed at a given depth.) Like Conshelf, it consisted of three tests:

>> **SeaLab I** was launched in 1964 at a depth of 59 meters (192 feet).

>> **SeaLab II** (see Figure 18-8) was launched in 1965 at a depth of 62 meters (205 feet).

>> **SeaLab III** was launched in 1969 at a depth of 190 meters (610 feet), after which the program was halted.

FIGURE 18-8:
SeaLab II.

Source: OAR/National Undersea Research Program (NURP); U.S. Navy, Public Domain

Aquarius

Aquarius is the only underwater, offshore, open water laboratory still in operation today. Now owned and operated by Florida International University, the 400-square-foot living space/laboratory sits about five and a half miles off Key Largo at a depth of 19 meters (63 feet), but the living space is actually at 15 meters (50 feet) below the surface. Because the station is at depth, the people staying on Aquarius, called Aquanauts, become "saturated," meaning after about a day, their bodies become saturated with dissolved gasses allowing them to live and work at this depth for days or weeks at a time, and they can dive deeper and for longer periods. Dives from Aquarius can last up to 9 hours as compared to a dive of similar depth starting from the surface, which lasts only about an hour before having to decompress.

Checking Out Other Ocean Monitoring Gadgets and Technologies

Scientists have far more in their toolbox for collecting data about the ocean than submersibles, sea labs, and diving gear. Devices are now operating 24 hours a day, 7 days a week, 365 days a year gathering data from all across the ocean and even from space to gain a deeper understanding of the ocean and to monitor changing conditions. In this section, we highlight some of the latest gadgets and technologies for studying and monitoring the ocean.

Buoys (moored and drifting)

Data buoys (also called *weather buoys*) are used all over the world to collect data such as wave height, wind and weather conditions, water temperature, water quality, pressure, and more. They're typically grouped in two categories:

>> *Moored buoys* are attached to the seafloor — sometimes 6,100 meters (20,000 feet) deep. They're used primarily to gather data that can be used to help predict weather and tsunamis. Figure 18-9 shows a NOAA Deep-ocean Assessment and Reporting of Tsunamis (DART) buoy.

» *Drifting buoys*, well . . . drift along. They're used mostly to study currents and temperatures. Some are also used to measure air pressure or surface salinity and to study how the ocean and atmosphere interact during storms, such as hurricanes.

Source: Stuart Hayes NOAA/NWS/NDBC, Public Domain

FIGURE 18-9:
A NOAA DART buoy.

Buoys are decked out with multiple instruments to take measurements and transmit data. Almost all are solar powered. On average a single buoy costs about $50,000 or more and lasts about 3 years.

FUN FACT

You can see the real-time data gathered from weather buoys all over the world at NOAA's National Data Buoy Center: www.ndbc.noaa.gov. Buoys also make great resting places for birds and pinnipeds (much to the chagrin of many buoy engineers).

Coring, dredging, and trawling tools

Geologists, biologists, and paleontologists have several methods for gathering samples from the ocean floor and several tools for each method:

>> **Coring** is used to collect sediment samples from the ocean floor to determine the physical characteristics of the sediment and study the layers of sediment to gain insight into Earth's past. Core samples contain a record of our planet, dating back millions of years, and they provide the keys for unlocking the mysteries of climate changes, mass extinctions, and evolution of life. Corers come in two types — box corers and tubular corers, each of which includes a number of subtypes. Think of them as cookie cutters or apple corers that are pushed down into the seafloor to collect a small cross-section of it. (See Chapter 6 for more about core samples.)

>> **Dredging** is used to collect loose rocks and other items from the ocean floor using a chain-link bag with a large metal-jawed opening that scoops items into the bag.

>> **Trawling** involves dragging a net through the water or along the seafloor to gather marine organisms at different depths, including mollusks, crustaceans, and fish.

Water column samplers

Scientists collect samples of water to understand the microscopic biology and the chemical characteristics at different locations in the ocean and at different depths.

The bottles used to collect samples aren't simply test tubes or glass jars for gathering water. They're equipped with advanced electronics that enable them to open and close to capture a sample at a preprogrammed depth.

To understand currents, scientists need information regarding the water's density and how it changes with depth. The density of water depends on its temperature, salinity (measured by the water's ability to conduct electricity), and depth. For this, scientists use conductivity, temperature, and depth (CTD) sensors, which can transmit the data back to the ship.

Entire networks of water column samplers are floating around the ocean collecting and transmitting data as we write this. For example, Argo is an international program that uses a fleet of robotic water column samplers constantly drifting with the ocean currents and moving up and down in the water column to collect and transmit data about water pressure, temperature, and salinity along with certain biological characteristics of the water. As of 2020, Argo's nearly 4,000 operational floats were collecting 12,000 data profiles monthly (400 daily) worldwide. Visit `argo.ucsd.edu/about` for more about the Argo program.

Sonar and lidar

Sonar (short for sound navigation and ranging) and *lidar* (short for light detection and ranging) are technologies used to map the ocean floor, as explained in Chapter 6, and to identify objects below the surface, such as shipwrecks. The technology works by emitting a sound wave or a focused beam of light, measuring the time it takes to bounce off an object and return to the source, and then calculating distance based on that travel time. Passive sonar is also used to simply "listen" for acoustic signals emitted from other sources, such as whales, dolphins, or submarines.

Sonar and lidar technology have advanced to a point at which the data collected can be used to render three-dimensional images of the seafloor and of objects below the surface. So cool.

The Global Ocean Observing System (GOOS)

The global ocean observing system (GOOS) is a real-time data collection system designed to provide insight into the physical state and biogeochemical profile of the ocean at all points around the world. Data is collected from ships, buoys (moored and drifting), water column samplers, tide gauges, and satellites. Together, this data provides a variety of information such as the following:

>> Water level measurements and forecasts

>> Locations and strengths of currents

» Wave heights and forecasts

» Sea ice measurements and coverage

» Rainfall measurements and forecasts, along with drought and flood predictions

» Maps and forecasts of harmful algae blooms (HABs, see Chapter 8 for details)

» Fish stock vulnerability assessments

» Weather- or climate-related disease forecasts

For additional information about GOOS, visit www.goosocean.org.

IN THIS CHAPTER

» Appreciating the ocean's contribution to our seafood diet

» Tapping the sea for fresh water, energy, and minerals

» Leaning on the ocean for shipping, travel, and tourism

» Exploring the ocean for new medications

» Reaping other ocean perks

Chapter **19**

Tapping the Ocean's Resources: The Blue Economy

very country has an economy — a system of production, distribution, trade, and consumption of goods and services. Likewise, the ocean has an economy, often called the Blue Economy, which encompasses the economic activities that create sustainable wealth from the ocean and its many resources. The concept of a Blue Economy encourages the sustainable use of the ocean and its resources for economic development, job creation, and improving people's lives, while at the same time preserving the ocean for future generations.

The Blue Economy covers maritime transportation, renewable energy, fisheries, tourism, climate change mitigation, and waste management, along with effective stewardship of the ocean. In this chapter, we bring you up to speed on the Blue Economy while doing our best not to totally nerd out on you, but the idea of leveraging the ocean's vast resources to solve many of the world's problems while also restoring the ocean's to abundance gets us incredibly excited.

Imagine increasing wealth and opportunities, alleviating world hunger, supplying virtually unlimited clean energy, improving the quality of our air and water, and making our shorelines safer, all while cleaning the ocean and restoring its beauty and bounty. It's possible, but to bring this vision to fruition, we need to start being smarter about how we use our vast human and ocean resources.

REMEMBER

The combined contribution of the ocean economy has recently been estimated to be in the range of US$1.5 to $3.0 *trillion* annually — roughly 3 to 5 percent of *global* gross domestic product (GDP). On the flip side, if our ocean continues to degrade at its current rate, the cost to the global economy is estimated at $428 billion per year by 2050 and nearly $2 trillion by the year 2100. In other words, instead of *gaining* $3 trillion, we'll be *losing* $2 trillion — that's a $5 trillion flip *per year!* And it's preventable.

Supplying the World's Seafood Diet

Fish is the largest traded food commodity in the world — the global seafood industry is worth a whopping $190 billion annually, and approximately three billion people around the world rely on seafood as their primary source of protein. Wild fish require no extra land or water to produce, and unlike livestock, they're not a major source of greenhouse gas emissions. The cost per pound to harvest fish from the sea is lower than that of farming beef, chicken, lamb, or pork.

Unfortunately, we are currently fishing ourselves out of fish. By some estimates, 85 to 90 percent of all fish stocks worldwide are either completely overfished or fished to capacity.

Reckless fishing activities are costing the global economy billions of dollars a year and robbing coastal communities of jobs and food now and in the future. Transitioning to sustainable fishing could boost the current dwindling global catch by 15 percent, feed millions more people, increase fishing profits by $75 billion *annually,* and leave 36 percent more fish in the sea.

Here are five ways to make ocean fisheries sustainable:

>> **Fish smarter.** Many fish species, such as tuna, become more fertile as they age. One large female can release the same number of eggs as 30 to 60 smaller females. Leaving the bigger (and usually older) female fish in the ocean to reproduce can significantly boost populations overall. Sportfishing and spearfishing can be especially harmful if they target the biggest (and sometimes the most fertile) fish, which is the opposite of smart.

>> **Reduce bycatch.** *Bycatch* is fish and other marine creatures caught unintentionally while targeting a specific species. It accounts for about 40 percent of all fish pulled from the sea — around 28.5 million metric tons per year. Great work is happening in this area, including smarter nets, better hooks, and even fishing during specific hours (like at night when species like birds are less active and less likely to be caught accidentally), but more can be done.

REMEMBER

Due to a lack of demand for certain species and other factors such as unreliable refrigeration and transportation, one out of every three fish pulled from the sea never makes it to someone's plate. Closing the loop on this waste would benefit the fishing industry while keeping more fish in the sea.

>> **Improve international cooperation.** Only about eight countries really fish the open seas, and usually they overfish. With better international cooperation and enforcement, large commercial fleets can be forced to fish smarter and more responsibly.

>> **Reduce the number of large fleets.** Large fleets owned by a small number of people and often subsidized by governments are responsible for much of the overfishing. Reducing the number of boats in these fleets and eliminating these subsidies would help to address overfishing while distributing the wealth of the sea to more people, especially back to the smaller artisanal fishermen.

>> **Create no-take marine reserves in key biodiverse sites around the world.** No-take reserves provide a place for marine populations to breed and flourish, giving their populations a chance to recover. And, because fish know no boundaries, they swim outside these reserves, providing more fish to catch. Fish less, catch more . . . sounds like a win-win.

See Chapter 21 for additional solutions to overfishing.

Harvesting plant life, too

Fish aren't the only commodity the ocean produces. Seaweed and algae (including kelp) are being grown and sold for multiple uses, including food and food additives, animal feed, nutritional supplements, beauty products, and more. Sustainably growing these plants not only helps to feed people while generating profits, but it also improves water and air quality, sequesters carbon, and creates a safe place for fish and other sea creatures to live and breed. On top of all that, many of these plants are regenerative, meaning they grow back without having to replant them after the harvest.

Growing our own supplies: Aquaculture and mariculture

Aquaculture (growing aquatic plants and animals on land) and *mariculture* (growing them in the ocean) involves raising fish, crustaceans, mollusks, aquatic plants, and other aquatic foods and products under controlled conditions. In the United States alone, aquaculture is a $1.5 billion industry, employing 1.7 million people and producing about 284 million kilograms (626 million pounds) of seafood products in 2018.

Done properly, aquaculture and mariculture can be used to not only grow a specific product but to also improve the ecosystem around it. For example, a company called Jewelmer cultures beautiful golden South Sea pearls in the waters around the Philippines, while its oysters filter the water. The clean water has brought back an incredible amount of *biomass* (total quantity of living things) and biodiversity to the area — fish, corals, marine mammals, birds, and more — while improving local seafood harvests. The clean water and healthy ecosystem have also strengthened their native mangroves and coral reefs, which serve as a natural storm surge barrier. When the storm surge from typhoon Haiyan devastated the coastlines of much of the Philippines in 2013, the pearl farms, protected in part by their healthy coast lines, reefs, and mangroves, suffered a fraction of the damage other areas experienced. Responsible pearl farming not only protects the environment but also provides a sustainable livelihood for rural coastal communities. It is a livelihood that benefits from a healthy environment and thus incentivizes local communities to protect and restore local ocean ecosystems.

REMEMBER

Business models built on the mission of profiting from the ocean and improving people's lives while restoring the environment is what the Blue Economy is all about.

Tapping the Sea as a Source for Fresh Water

Only about 3 percent of Earth's water is fresh water, and two-thirds of that is locked in glaciers or other places people can't access it. As a result, 1.1 billion people lack access to fresh water, and another 1.6 billion have a scarcity of water for at least one month every year. Certainly the ocean has plenty of water to go around, but it's salt water — you can't drink it or water your plants with it.

Enter *desalination* — literally removing the salt from seawater. Desalinating seawater could provide for all the world's water needs . . . if it weren't for two big drawbacks — the massive amount of power consumed by desalinization plants and all the salt (brine waste) left behind in the process.

While awesome advances in renewable energy are reducing the need for fossil fuels to run these plants, the salt remains a major issue. If pumped back into the ocean, the extra salt increases the salinity of the water. Remember, even a slight increase in salinity can wreck entire ecosystems. Chemists and others are working on environmentally friendly ways to discard or, better yet, use the extra salt. For example, Farid Benyahia, a chemical engineer at Qatar University, has patented a process for converting the brine waste into sodium bicarbonate (baking soda) using carbon dioxide and ammonia (which is reclaimed in the process). This solution not only creates a usable product from the brine waste but also helps reduce the cost of carbon dioxide storage. Other solutions, anyone? Desalination plant table salt? Salt-based bio fuel? Houses built with salt blocks? Just spitballing here.

Shipping Goods 'Round the World

Ninety percent of everything we buy arrives on a ship, from cars to computers, to clothes, to food, and even most Bibles (China is the largest printer of the blessed book worldwide). And at any given moment, at least 20 million shipping containers are traveling across the ocean. (*What?!*) Seriously, the ocean shipping industry is crazy interesting (and just plain crazy). You can read all about it in the book *Ninety Percent of Everything*, by Rose George (Metropolitan Books). With the growth of the global economy, ocean freightliner traffic is expected to increase significantly.

The good news is that shipping freight by sea isn't quite as bad for the environment as shipping by air or land (it's the least of the three evils). Because the world will increasingly rely on the ocean to carry its freight, the greener and more efficient this industry is, the better it is for all of us. Interesting work is happening in this field, such as the development of cleaner fuels and hulls that move through the water with less drag, but there is huge room for improvement.

Digging Up Gold, Diamonds, and Other Valuables: Deep-Sea Mining

The ocean floor holds many treasures — by some estimates $17 billion to $60 billion in sunken treasures alone (gold, silver, and jewels that went down with their ships), but that's a penny in the wishing well compared to the amount of copper, cobalt, and nickel that may be resting on or below the surface of the seafloor. With our increasing reliance on smartphones, computers, solar panels, and batteries, the demand for these precious and rare-earth metals is on the rise, and so is their value.

Mining companies are stopping at nothing to get these rare materials. In places like the Congo, the land, along with the local people, are being devastated as unchecked greed fuels war, genocide, and even terrorism.

WARNING

With many of these precious elements available in the ocean, companies are racing to get their hands on them, potentially at a huge cost we don't yet fully grasp. For example, precious metals have been found in deep-sea thermal vents, and we have no way to predict the impact that mining would have on these vents and the ecosystems that develop around them. How much biodiversity is likely to be lost? What are the unintended consequences we aren't even thinking about?

Currently, most deep-sea mining isn't regulated. The fear is that mining companies will likely drag the seafloor, willfully destroying entire ecosystems in the pursuit of riches. And, according to the Law of the Sea (see Chapter 20), those riches belong to humanity — not to any one company. The sea could provide the raw materials to meet the world's future needs, but people need to proceed with extreme caution. Indeed, a better approach would be to invest in innovation that would eliminate the need for materials, so we don't have to extract deep-sea minerals in the first place. No matter what, more work needs to be done. We need to invest in more research, establish transparent regulation that puts the precautionary principle first, and determine protocols where any wealth gained is distributed equitably to the world before any action is taken.

Harnessing the Ocean's Energy Resources

As we explain in Chapter 16, the ocean is constantly in motion and absorbing heat from the sun. Tides, waves, currents, and wind represent a massive amount of mechanical energy, and relatively warm ocean waters provide additional thermal energy just waiting to be converted into electrical energy. According to the International Energy Agency, harnessing all this ocean energy could produce 20,000 to 80,000 terawatt hours (TWh) of electricity annually. That's more than enough power to meet the world's current total energy consumption of almost 20,000 TWh. Holy Terawattoli!

Obviously, harnessing all the ocean's energy isn't possible or even advisable, but the recoverable energy just along the U.S. coastline is estimated at 1,170 TWh annually, which is a third of the country's energy consumption. Not too shabby! Here is a quick list of just a few of the technologies that have the potential to be implemented to capture energy from the ocean:

» **Offshore wind energy:** Of all ocean energy, offshore wind holds the most promise and is already generating significant amounts of energy around the world. It is actually more promising than land-based wind. Why? Because wind blows more consistently across the ocean where nothing is in the way to slow it down, such as trees and mountains. So while the other ocean energy may be harnessed, offshore wind is already up and running (or should we say turning?) and generating clean energy for us.

» **Wave energy:** Several devices have been designed to convert the mechanical action of waves into electrical energy. One such device is a point absorber buoy anchored to the seafloor. As the buoy moves up and down it moves a generator that creates electricity. Other devices use various methods to produce a pressurized flow of water that drives a turbine that generates electricity.

» **Tide energy:** Tides move massive amounts of water, and this water movement can be used to generate electricity. One device involves a dam-like structure with gates through which water flows in and out to turn turbines that generate electricity.

» **Thermal energy:** The temperature difference between warm surface water and cold deep water can be used with the addition of heat exchangers to turn turbines that generate electricity (also known as OTEC).

» **Salinity gradient power:** Salinity gradient power (also known as osmotic power) is based on the fact that when two solutions of different concentrations of salt are separated by a semipermeable membrane, the molecules of the less concentrated solution (fresh water) move to the area of higher concentration (salt water). This movement can be used to turn a turbine that generates electricity.

PUMPING OIL AND GAS FROM BELOW THE SEAFLOOR

Plenty of oil and gas reserves are available below the seafloor, but drilling and pumping at the bottom of the ocean is risky business. It's not a question of whether another disaster will occur but when. Massive cleanups continue in the aftermath of the 2010 (BP) Deepwater Horizon oil spill and the 1989 (Exxon) Valdez oil spill, and oil even still lingers in the environment from the 1979 (Pemex) oil spill that happened off the gulf coast of northern Mexico.

The ecosystems in the Gulf of Mexico are still suffering from the consequences of the 2010 Deepwater Horizon oil spill long after the company's executives stopped writing checks to reimburse the people most affected. To add insult to injury, most oil companies are permitted to write off the cost of their oil spills on their taxes (seriously, look it up).

The best solution to deep-sea oil disasters is to stop the drilling and invest in renewable energy sources instead. The threat they pose to wildlife, people, the economy and future generations is too great. Besides, do we really need more fossil fuels to burn when we're already suffering from the catastrophic effects of climate change?

Discovering New Medications

Many of our medications come from nature — historically from terrestrial plants and organisms. However, medical researchers are increasingly turning to the sea for cures and treatments. Over the past 30 years, they've extracted at least 20,000 new biochemical substances from marine creatures. Currently in use are an anti-tumor drug derived from sea-squirts and a painkiller formulated from cone snails, not to mention the HIV drug AZT which was derived from a shallow water sponge found in the Caribbean. And as bacteria continue to become more resistant to current antibiotics, scientists are turning to the sea for answers.

Medical research like this is possible only when the ocean and its ecosystems are healthy. If we humans fall short of protecting and preserving the ocean and its inhabitants, such research will grind to a halt. Again, a healthy ocean means healthy people . . . in this case, quite literally.

Capitalizing on Tourism and Recreation

Dear ocean, how do I love thee, let me count the ways. Seaside vacations, swimming, diving, snorkeling, boating/sailing, personal watercraft, recreational fishing, surfing, wind-surfing, whale watching, shell collecting, and more. The ocean is *the* hot spot for fun, relaxation, and exploration, and it supports a booming tourism industry.

According to research by The Nature Conservancy, the World Bank, and economists, ocean-based tourism and recreation in the United States alone employs almost 2.5 million people collectively earning a total of about $58.7 billion annually, and it contributes about $124 billion to GDP each year. The coral reef tourism sector alone is valued at an estimated $36 billion. And more than 70 countries and territories around the world have "million dollar reefs" — reefs that generate over $1 million in tourism spending annually. What do these reefs ask for in return? Nothing, except for us to not destroy them . . . is that too much to ask?

FUN FACT

Even better, tourism helps spread the wealth. Tourism is the largest voluntary transfer of wealth in the history of humanity.

Accounting for a Few Ancillary Ocean Benefits

The ocean showers us with blessings, many of which benefit everyone without really profiting anyone. We mention many of them throughout this book directly or indirectly; for example, in Chapter 16, we highlight the importance of ocean currents in distributing heat and nutrients around the world, and in Chapter 17, we explain the ocean's role in controlling climate and generating weather systems. In this section, we showcase a few additional ways the ocean benefits all of us that extend beyond the Blue Economy.

Carbon storage (a.k.a. blue carbon)

The ocean is one of our most powerful weapons for combating climate change because it has absorbed about a quarter of the anthropogenic (human-caused) carbon dioxide, CO_2, produced by fossil-fuel combustion and deforestation. For comparison, forests on land also absorb about a quarter of our CO_2 emissions. The half of our emissions that is not absorbed by the ocean and land plants accumulates in the atmosphere and, along with other man-made greenhouse gases, is causing the planet to warm.

The way the ocean takes up (or *sequesters*) man-made CO_2 is related to the ocean's global conveyor belt (see Chapter 16). When humans produce CO_2 by burning coal, oil, and natural gas or cutting down trees, the CO_2 produced quickly dissolves in the ocean's surface layer (causing a decrease in seawater pH; see Chapter 3). If this were the whole story, the ocean's ability to take up CO_2 would not be nearly as large as it actually is. Cold and salty surface water off the coast of Greenland and Antarctica is dense enough to sink to the bottom. As it does so, it carries the extra man-made CO_2 to the abyss where it will stay for about a thousand years.

The ocean also sequesters *blue carbon,* organic carbon that's collected and stored in coastal marine ecosystems such as mangroves, seagrass beds, and tidal marshes. While these unique coastal ecosystems account for only about 0.2 percent of the total ocean, they're responsible for about 50 percent of the carbon sequestered in ocean sediment, and some experts think that number is as high as 70 percent!

What is the economic value of these marine plants? Well, mangroves are estimated to be worth at least US$1.6 billion each year in ecosystem services that support coastal livelihoods and human populations around the world. And that's just mangroves. Imagine how much seagrass and tidal marshes would add to that number. Like many animal species (such as sharks) these healthy ecosystems are worth far more alive than dead, and sadly, for a long time these vital ecosystems were disappearing faster than our rainforests, even though rainforests hogged the headlines. (We like a good rainforest as much as the next person, but let's give credit where credit is due.) Fortunately, increasing recognition of their value is slowing down the losses and even reversing them in some regions.

But more than sequestering carbon in plants, we can also sequester carbon in animals. As our good friend and leading marine biologist Carlos Duarte summarizes: "Today, there is only 20 percent of the population of great whales left that roamed the ocean before whaling started. Now remember that we did not hunt them for their meat, but for their oil. We literally burned these whales to light the streets of Europe, North America, and other regions. What once was carbon trapped in whales was burned and let loose as CO_2 in the atmosphere contributing to the warming of our planet. But what could happen if we rebuilt marine life, restoring populations of great whales and all the rest of the large ocean animals that we have decimated? Imagine turning the carbon that is loose in the atmosphere, once again, into a magnificent blue whale, not only keeping the oceans productive but catalyzing and enhancing the ocean's power to remove CO_2 from the atmosphere. Rebuilding marine life by 2050, a doable goal, decarbonizes the atmosphere to recarbonize the biosphere in the form of the countless extraordinary marine life that exists."

For more on this exciting concept, check out the following article by Carlos and his colleagues: 2020. "Rebuilding Marine Life." *Nature* 580: 39–51, https://doi.org/10.1038/s41586-020-2146-7.

Coastal protection

Healthy reefs, mangroves, barrier islands, seagrass meadows, and sand dunes are like shock absorbers, buffering the impact of storms, storm surges, and tsunamis on coastlines around the world (see Figure 19-1). Anyone living on or near a coastline has these valuable ocean buffers to thank for protecting them, their land, and their property. And all that we humans need to do is let these storm busters do their jobs.

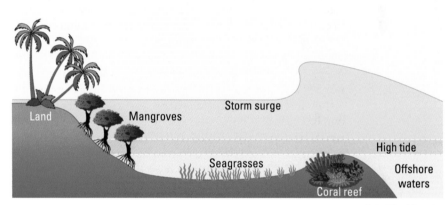

FIGURE 19-1:
Reefs, seagrass meadows, and mangroves provide a buffer against storms.

©John Wiley & Sons, Inc.

Coral reefs serve as the first line of defense, absorbing up to 97 percent of the wave energy and protecting 63 million people around the globe. A bank of mangroves 100 meters (330 feet) thick can reduce wave heights between 13 and 66 percent, while a bank 500 meters (1,640 feet) thick can stop waves in their tracks. In the Philippines, Mexico, and Malaysia, the annual coastal protection benefits from reefs exceed $450 million . . . for *each* country. And oyster reefs can save communities $85,000 each year per hectare if used in place of artificial breakwaters.

Cultural value

To many coastal communities, the ocean plays an integral role in their cultures. These cultures have developed in and around the ocean, and have come to rely on it for food, clothing, shelter, and customs. Over the course of hundreds or thousands of years, they've developed belief systems, ceremonies, dances, art, and narratives to capture and celebrate their integral relationship with the sea.

In terms of practical value, these same communities have developed best practices and protocols for the sustainable management and development of ocean resources that can teach us all a lesson or two.

Fortunately, ethnoecology experts and others are collaborating with indigenous coastal communities to identify and restore traditional marine management systems and practices. Their hope is to not only restore and enhance marine ecosystems, but also to provide these cultures with the means to maintain their communities and their rich cultural heritage.

Biodiversity

When you visit a thriving ocean ecosystem such as a coral reef or you watch nature videos showcasing the incredible diversity of marine life, you can't help but to feel awed by nature's splendor. The beauty alone is enough to make most people realize that preserving such biodiversity is of the utmost importance. After all, who wants to live in a world where only a few species dominate the environment?

However, the aesthetic value of biodiversity is just one component of its value. Biodiversity is critical to the survival of the ocean and the planet as we know it. Even the loss of a single species, or the introduction of a non-native species, can destroy an ecosystem. And with the loss of ecosystems and the decline of biodiversity, we lose many of the benefits of the Blue Economy, including tourism, fisheries, clean water and air, new medicines, and opportunities to learn and discover more about the ocean and ourselves. To learn more about restoring our ocean to abundance, check out Alexandra Cousteau's initiative (Philippe's sister) at https://www.oceans2050.com and for more of a deep dive on the economic value of our ocean, visit oceanwealth.org.

REMEMBER

Our leaders in government and business, and the rest of us for that matter, need to wake up to the fact that the ocean is life and livelihood for all of us, plain and simple.

Chapter **20**

Governing the Ocean: Treaties, Laws, Agreements, and Enforcement

E xcept for a few landlocked countries, most countries have some beach-front property, and all countries have an interest in how the ocean and its resources are managed. To ensure the safety and security of vessels, including their crews, passengers, and cargo; to provide equitable distribution of the ocean's resources; and to preserve the ocean and its wildlife for future generations, countries around the world have enacted laws and entered into international agreements.

In this chapter, we take a look at some of the treaties, laws, and agreements in place to govern the use of the ocean and its resources and explain the overall systems most countries have in place to enforce the rules.

Recognizing the Two Systems of Law That Govern the Seas

Two systems of law govern behaviors and actions in ocean waters: Admiralty Law (otherwise known as Maritime Law) and the Law of the Sea. Admiralty Law generally applies to navigation and shipping for the purpose of facilitating commerce among private parties. The Law of the Sea, on the other hand, consists of customs, treaties, and international agreements for maintaining peace, order, and productivity in ocean waters among countries. In the following sections, we explain these two systems of law in a little more detail.

Admiralty Law

Admiralty Law governs nautical issues and private maritime (commercial or military) disputes involving private parties using ocean-going vessels. It addresses issues such as seafarers' rights, salvage operations, mortgages, pollution, navigation, commerce, care of passengers, and maintenance and cure (the equivalent of workers compensation — *maintenance* covers an employee's daily living expenses, and *cure* covers medical costs).

While different countries have different laws, the United Nations (UN) created the International Maritime Organization (IMO), which has established a generally accepted set of rules adopted by many countries. Those rules, however, are too numerous and complex to go into detail here. But if you are curious, check out more information about the IMO at imo.org. The UN also provides dispute resolution, assuming the parties are signatories to the Law of the Sea (discussed in the next section).

FUN FACT

Admiralty Law goes waaaaay back. Early versions were practiced by the Romans and Phoenicians. However, the modern concept of Admiralty Law is thought to date back to the French Queen Eleanor of Aquitaine in the late 12th century. She was married to King Louis VII and joined him on a crusade to the Middle East, where she first was exposed to the idea of a maritime law separate from common law. She brought this concept home and instituted it on the island of Oleron, in the form of a law known as the Rolls of Oleron, to govern maritime issues in her territory in France. She then brought it to England when she was serving as regent for her son King Richard the Lionheart (remember him from Robin Hood?). Later, the British brought Admiralty Law to the Colonies, which later became the United States of America.

The Law of the Sea

The 1982 United Nations Convention on the Law of the Sea (UNCLOS) is best described as the primary playbook governing how the ocean can be used on a global scale. It entered into force in November 1994. Commonly referred to as The Law of the Sea Convention, it establishes a legal framework for both the use of marine resources as well as their protection. It covers all areas from related airspace, down through the water column to the seabed and even the sub-seabed. It governs everything from seabed mining to marine conservation to dispute settlement to how the ocean is divvied up.

Key provisions of the Law of the Sea address the following areas:

» **Setting limits:** What nations can do and where they can do it, which includes the definition of *territorial waters* (explained in the next section) and *innocent passage* (which allows for a foreign vessel to pass through another country's territorial waters subject to certain conditions).

» **Navigation:** Establishing where ships can and cannot go, for example, establishing freedom of navigation in the high seas and determining how certain narrow straights (such as the strait of Gibraltar), which are technically within territorial waters but are important to all nations, are governed.

» **Exclusive economic zone (EEZ):** An *exclusive economic zone* is an area of coastal water and seabed within a certain distance from a country's coastline in which the country claims exclusive rights to fish, drill, mine, and conduct other economic activities. The Law of the Sea stipulates how these zones are defined and the activities permitted.

» **Continental shelf:** A *continental shelf* is an underwater extension of a continent, which is shallower and therefore more easily accessible for commercial exploitation. The Law of the Sea defines "continental shelf" and stipulates the extent to which a nation can claim rights to it. Understandably, nations want to claim as much continental shelf as possible, so restrictions needed to be established.

» **Deep seabed mining:** The Law of the Sea declares resources of the seabed (which can be rich in minerals and precious metals; see Chapters 19 and 21) beyond the reach of any nation and "the common heritage of mankind," and therefore under the UN's control.

» **The exploitation regime:** The Law of the Sea established a group within the UN, The International Seabed Authority, to manage the exploitation of seabed resources.

» **Technological prospects:** The Law of the Sea includes some insights into the technological means of deep seabed mining.

- **The question of universal participation:** Establishes a way for countries to join the convention and determines what happens to those countries that choose not to join. The U.S., for example, has still not ratified the Law of the Sea despite the fact that historically both Republican and Democratic presidents have supported it, as have NOAA, the Coast Guard, NASA, the Pentagon, and the majority of senators and congressional representatives. Regardless, a few hardcore conservative politicians led by Senator James Inhofe (who has one of the worst environmental track records in history) have consistently blocked it, claiming distrust of international treaties. The problem is, if you don't have a seat at the table, you can't influence the outcome.

- **Pioneer investors:** Protects the interests of those nations that were the first to invest in seabed mining technologies.

- **Protection of the marine environment:** The Law of the Sea covers several different types of marine conservation, including land-based and coastal activities; continental-shelf drilling; potential seabed mining; ocean dumping; vessel-source pollution; and pollution from or through the atmosphere. It also seeks to regulate and protect the sustainability of fisheries on the high seas and resolve disputes between members about resource exploitation, stating that it's the fundamental obligation of all states to protect and preserve the marine environment.

- **Marine scientific research:** Deals with the rights of nations to conduct research within their EEZs as well as the territorial waters of other nations. It stipulates that while nations must ask for permission first, said permission will not be withheld "in normal circumstances" and "shall not be delayed or denied unreasonably." Crucially, if permission is requested and the state from which permission was requested doesn't reply within six months, it's deemed that permission was granted by default.

- **Settlement of disputes:** This lays out how to settle disputes between members. It gets really legalese and confusing very quickly so we won't go into it here but suffice it to say, it encourages nations to work together and establishes international arbitration rules if they can't.

For more about UNCLOS, visit www.un.org/Depts/los/convention_agreements/convention_overview_convention.htm.

FUN FACT

The need for a framework to govern how the ocean is used has always been an issue of international concern, because throughout history, some countries have made extravagant claims. For example, two years after Christopher Columbus first visited what would soon be called America, Pope Alexander VI had a meeting with officials from the two biggest maritime powers of the time — Portugal and Spain. The purpose of the meeting was to divide the Atlantic Ocean in two. Everything west of a line the Pope drew down the middle was given to Spain, and everything east was given to Portugal. So, by that measure, Spain controlled the Gulf of Mexico and the Pacific Ocean (like the entire thing) and Portugal got the South Atlantic and the entire Indian Ocean. What about everyone else, including the people who lived there already? The Pope, along with the leaders of Spain and Portugal (of course), didn't care, often citing God as granting them superiority over "the locals." Colonialism — you gotta love it (NOT!).

Establishing Sovereign and International Jurisdictions

Critical to any understanding of the Law of the Sea is a recognition of the different areas of the ocean and the activities permitted and prohibited in each area. The Law of the Sea essentially establishes two jurisdictions:

» **Sovereign jurisdiction:** The area of the ocean and seafloor controlled by a country

» **International jurisdiction:** Areas of the ocean outside the control of any individual country

Understanding sovereign jurisdiction

Every coastal nation has what's considered *territorial water,* which extends out to 22 kilometers or 12 nautical miles (note that a *nautical mile* is slightly less than a mile on land). This area is considered within the exclusive sovereign control of a nation both above and below the surface. Within this zone, a coastal nation is granted jurisdiction over the exploration and exploitation of all marine resources, such as energy production, mining, and fishing. The only exception is a term called "innocent passage," which allows a vessel from any nation to pass through another's territorial waters as long as they don't engage in any activity that exploits those waters, such as weapons testing, spying, smuggling, severe pollution, fishing, or scientific research.

When a nation comprises several states (such as the United States), each state claims its own territory generally between 5 and 19 kilometers (3 to 12 miles), beyond which the nation has its sovereign jurisdiction.

FUN FACT

The 12 nautical mile limit wasn't formally established until 1982. Prior to the 19th century, nations generally agreed to a distance of 5.5 kilometers (about 3 nautical miles). Why this distance? Because, at the time, it was the range of a canon shot.

In addition to territorial waters, any body of water that's partially or fully enclosed by land, such as lakes, bays, or waters encompassed by an archipelago of islands (think of the Philippines or Indonesia) is also considered territorial waters.

Extending sovereignty across contiguous zones

The Law of the Sea also allows coastal states to claim a *contiguous zone* that extends up to 44 kilometers (24 nautical miles) from its coast, where the nation state is permitted to exercise the control necessary to "prevent the infringement of its customs, fiscal, immigration or sanitary laws and regulations within its territory or territorial sea, and punish infringement of those laws and regulations committed within its territory or territorial sea." Think of the contiguous zone as a buffer zone that enables a nation state to more effectively protect its territorial waters.

Extending sovereignty to exclusive economic zones (EEZs)

Exclusive economic zones (EEZs) are waters and seafloor that a nation claims out to a distance of up to 370 kilometers (200 nautical miles) offshore (or less if a nation is closer than 741 kilometers [400 nautical miles] to another). This concept for maritime governance was adopted at the United Nations Convention on the Law of the Sea in 1982 and includes control and exploitation of only those resources *below* the surface.

Addressing the gray areas

Seems pretty straightforward, right? Coastal nations can claim up to 22 kilometers (12 nautical miles) off the coastline as territorial waters, a contiguous zone up to 44 kilometers (24 nautical miles) off the coastline, and an exclusive economic zone up to 370 kilometers (200 nautical miles) off the coastline. However, clarity fades when countries are closer than these maximums. Consider the following gray areas:

>> **Countries are closer than 741 kilometers (400 nautical miles) from one another.** The obvious solution is for two countries to split the difference; for example, if two countries are 556 kilometers (300 nautical miles) apart, each claims an EEZ of 278 kilometers (150 nautical miles) out from its coast. However, that's not always an equitable split. Countries often dispute this equal distance measure based on any number of reasons from fisheries concentrations to the location of undersea resources to geopolitical concerns.

>> **A country claims a body of water (such as a Gulf) that extends beyond 44 kilometers (24 nautical miles) from its borders as its territorial waters.** For example, Libya claimed the entire Gulf of Sidra as its territorial waters, effectively

extending its territorial waters to more than 426 kilometers (230 nautical miles) offshore. This claim was disputed by the United States and repeatedly challenged in the 1980s.

>> **Multiple countries claim certain islands or other areas of land as their territory.** For example, the Spratly Islands in the South China Sea are a significant source of tension in the area. They consist of hundreds of islands, cays, and reefs that represent around 2 square kilometers (0.7 square miles) of land mass scattered across an area of 425,000 square kilometers (164,000 square miles). The problem is that these islands are close to the borders of several countries and contain rich fishing grounds, strategic military locations, and key shipping lanes. All or part of these islands are claimed by several countries, including Vietnam, Malaysia, Brunei, Taiwan (ROC), China (PRC), and the Philippines.

>> **Countries build land masses in the ocean and then claim ownership of them.** For example, China has built a series of artificial islands in the South China Sea on top of atolls and even submerged coral reefs to try to reinforce their claim that most of the South China Sea is their territory. Depending on whose side you're on, this move constitutes ingenuity or cheating on the part of the Chinese government. Yeah, *not* cool, Dude.

REMEMBER

Coastal nations are highly motivated to extend their territorial waters and EEZs as far out as possible, so China's desire to control the South China Sea is understandable even if it is very, very unneighborly. Currently, the South China Sea is considered high seas and thus exploited by many nations. If any nation is allowed to claim it as part of its EEZ, it gets exclusive access to those areas, providing that nation with a huge economic windfall as well as strategic military assets. According to information gathered by the council on Foreign Relations, more than $5.3 trillion worth of shipping travels through the South China Sea each year, $1.2 trillion of which belongs to the United States. In addition, the South China Sea is thought to have enormous oil and natural gas reserves, and it accounts for about 10 percent of the global fish catch. So you can see why everyone wants a piece.

Ruling the high seas: International jurisdiction

In essence, any part of the ocean that's not claimed as part of a country's EEZ is part of the high seas and is owned by no one. The concept of "freedom of the high seas," which for a long time was generally accepted in practice, is now accepted law and is recognized to include freedom of navigation, fishing, the laying of submarine cables and pipelines, and overflight of aircraft. Governance of the high seas, like other areas of the ocean, is achieved through an alphabet soup of different conventions, laws, treaties, and regulations too extensive to cover here.

Policing the Ocean and Enforcing the Laws

Enforcement of Admiralty Law and the Law of the Sea falls primarily on the shoulders of coastal nations. Although each nation has its own military and law enforcement agencies called by different names, most countries have three distinct groups in charge of protecting their maritime interests: one that is involved in international matters, a second that patrols and protects their domestic waters, and usually a third that's a combination of merchant/private individuals who serve a variety of other tasks and support roles. In the United States, for example, the following three groups are in charge of policing international and domestic waters:

>> **The United States Navy:** The mission of the Navy is to maintain, train, and equip combat-ready Naval forces capable of winning wars, deterring aggression, and maintaining freedom of the seas. The Navy operates mostly internationally, but exceptions do occur at times.

>> **The United States Coast Guard:** The mission of the United States Coast Guard is to ensure the nation's maritime safety, security, and stewardship. The Coast Guard is a domestic force, but again, exceptions are made when circumstances call for them. Missions are broken down into two categories:

   - Homeland security, which is in charge of ports, waterways, and coastal security; drug interdiction; migrant interdiction; defense readiness; and other law enforcement

   - Non-homeland security, which is tasked with marine safety; search and rescue; aids to navigation; living marine resources; marine environmental protection; and ice operations

>> **Merchant Marine:** The least well known of the three maritime services, the Merchant Marine is a federally managed service separate from the military. It consists of civilian mariners and civilian or federally owned ships that engage in commerce or the transportation of goods in and out of U.S. waters. The Merchant Marine is associated with a variety of vessels, including deep-sea merchant (cargo) ships, ferries, dredges, tugboats, and excursion vessels. While not technically in charge of policing either domestic or international waters, it does operate ships that support the Navy, such as fuel re-supply vessels. The United States Maritime Administration, under the Department of Transportation, finances and administers programs that train Merchant Mariners.

Many countries follow this same or a similar breakdown when it comes to protecting and policing their waters.

Preventing illegal, unregulated, and unreported (IUU) fishing

Illegal, Unregulated, Unreported (IUU) fishing is one of the most serious threats to the ocean. It undermines all efforts to manage fisheries sustainably and often robs developing countries (which have the fewest resources to combat it) of precious resources that feed people and would otherwise help to develop their economy. According to at least one estimate, 20 percent of the fish sold in the United States is from IUU sources. Better international cooperation is needed to enforce the rules on the high seas, as well as investment in technology that can catch the bad guys in the act. It will also require a recognition by nations around the world that IUU is a threat to our security, our health, and our economy, as well as the environment, and act accordingly.

WARNING

IUU fishing threatens food security, increases poverty, and can even destabilize entire regions as exemplified with the emergence of the Somali pirates (see the nearby sidebar). IUU also leads to a great deal of other crimes like human slavery . . . seriously. To find out how, read Ian Urbina's excellent book, *Outlaw Ocean* (Alfred A. Knopf).

Combating piracy on the high seas

While images of Johnny Depp's quirky Captain Jack Sparrow may be what comes to mind when you hear the world "pirate," the truth is far more unpleasant. Piracy on the seas has been a problem for as long as people have boarded boats and headed off into the ocean. While we may think of swashbuckling pirates of the 1700s when we hear the word, today, piracy still presents a very real and deadly risk around the world. In fact, over the past decade piracy has increased, so much so that the United Nations Office on Drugs and Crime has established a Global Maritime Crime Program to combat it. However, as explained in the nearby sidebar, sometimes combating crime on the high seas might be more effectively accomplished through conservation efforts and some high-tech tools.

FUN FACT

While white European men like Black Beard and Calico Jack dominate the popular image of pirates, arguably the most successful pirate of all time was not European, male, or white. She was called Ching Shih and terrorized the seas around southeast Asia during the 19th century. At the height of her power she is thought to have commanded up to 18,000 ships with up to 80,000 men under her control. She struck fear into the hearts of Portuguese, British, and Chinese navies. After years as a successful pirate, a concerted effort by these nations finally convinced her to surrender. She negotiated a pardon for herself and most of her thousands of crew and retired to Macau to live out her days in comfort and fame. #Boss!!!

SOMALI PIRATES

In 1991, Somalia was hit by a civil war that collapsed its government, leaving the country without a serviceable coast guard to defend its waters. These were fertile waters that had supported small-scale local fisherman for generations. All of a sudden, huge foreign fishing vessels were coming to the Somali coast and hauling out everything their nets could hold, illegally. It's estimated that these illegal operations were stripping $300 million worth of fish every year out of Somali waters. This left the Somali fishermen unable to even feed their own families. So what did they do? They armed themselves to protect their traditional fishing areas and started capturing the illegal fishing boats. Not wanting to lose their high-priced fishing vessels, not to mention their crews, the owners of the captured vessels offered ransom for the safe return of their property and personnel.

Lightbulbs flashed in the minds of the Somali fishermen, and a whole new industry was born. Terrorist organizations like Al Shabbab and Al Qaeda in the Maghreb got involved and armed the fishermen in exchange for a cut of their profits, thereby helping to fund terrorism all over the world. As a result, the global economy has lost countless billions and invested hundreds of billions more in military interdiction not to mention the human cost: Thousands of people have died as a result of this terrorism, and hundreds of thousands more have been displaced.

In a summary of a U.S. Department of Defense report on the crisis, the last line reads, "as described, simply spending a few million dollars on protecting fish habitats could have prevented Violent Extremist Groups (VEOs) from metastasizing in east Africa, costing more lives and billions in treasure."

Think about that . . . a few million dollars of conservation could have saved countless lives and hundreds of billions of dollars while preventing untold suffering.

Chapter **21**

Taking Care of the Ocean That Takes Care of Us

To think of the ocean as an unlimited resource is tempting. After all, it covers more than 70 percent of Earth's surface, has an average depth of 3,720 meters (12,200 feet), and contains about 1.3 billion cubic kilometers (320 million cubic miles) of water, accounting for about 97 percent of the total amount of water on the planet. Running out of ocean resources is as difficult to imagine as running out of air to breathe.

However, today Earth is also home to more than 7.5 billion people who consume resources and produce waste, much of which eventually ends up in the ocean. In 1825, when the world had only about one billion people, no automobiles, and little inorganic and toxic waste, the human impact on the oceans was negligible. In the relatively short period of 200 years, the human population has nearly doubled three times, global carbon emissions have increased 20-fold, production of plastics increased 200-fold (from 1907 when the first plastic was invented), global pesticide production increased from zero to more than three million metric tons (from the 1940s to 2000), and total seafood catch from industrial fishing alone quadrupled from less than 20 million to more than 80 million metric tons (from 1950 to 2000).

Most experts agree that the ocean is at its tipping point, and the next ten years are critical in restoring and preserving the ocean for future generations so our children can enjoy its wonder and majesty (see Figure 21-1). In this chapter, we identify the most serious problems and possible solutions while highlighting a few ecological success stories.

FIGURE 21-1: Antarctic sea ice. Just one example of the beauty and majesty of the ocean.

Source: John Weller – www.sealegacy.org

Keeping Tabs on Ocean Health

Scientists and conservation groups around the world monitor the health of the ocean and coastal areas in different ways to observe and record changes over time. One such group publishes the *Ocean Health Index* (OHI), which compiles scores from 220 *Exclusive Economic Zones* (EEZs) around the world to determine a global OHI. The study assesses "the marine waters under national jurisdiction, the coastlines, and oceans (out to 200 nautical miles), of 220 coastal countries and territories. These regions total 40% of the ocean and provide most benefits to people, but also incur the most pressures from human activities."

Scores range from 0 to 100, with 100 representing a "healthy ocean," meaning a sustainable human-ocean ecosystem that delivers a range of sustainable benefits to humans without harming the ocean's web of life. EEZs were ranked in ten categories framed as goals:

- » Food provision (wild-caught fisheries and *mariculture* [seafood farm-raised in the ocean or along the coast])

- » Artisanal fishing opportunities (small-scale local fishing, as opposed to large-scale industrial fishing)

- » Natural products (coral, fish oil, ornamental fish, seaweed and plants, shells, and sponges)

- » Carbon storage (plants and sediment in natural coastal ecosystems that capture and store large amounts of carbon)

- » Coastal protection (habitats that protect shorelines from waves and flooding, including mangrove forests, seagrass meadows, salt marshes, tropical coral reefs, and sea ice)

- » Sense of place (the cultural, spiritual, aesthetic, and other intangible benefits of the ocean as measured by iconic species and lasting special places)

- » Costal livelihoods and economies (the people and communities that rely on the oceans to live)

- » Tourism and recreation (the percentage of a country's total labor force engaged in coastal or marine tourism)

- » Clean waters (absence of chemicals, excessive nutrients, human pathogens, and trash)

- » Biodiversity (the variety of marine life)

The 2019 Global OHI was 71, which indicated a slight improvement over 2018. According to the OHI team, the 2019 score "sends a message that the ocean isn't 'dying' as many people think," but there is considerable room for improvement.

Zeroing in on the Problems

The ocean is facing many issues, some big, some bigger. Some of these issues are very hard to tackle because they're complex and interrelated and because the ocean system takes a long time to recover.

Think of an unhealthy ocean as an unhealthy person. If a healthy, fit individual gets an infection, such as pneumonia, she has a good chance of kicking it and recovering fairly quickly. In contrast, when someone with a chronic illness, such as diabetes or high blood pressure, or someone who's been smoking for many years gets pneumonia, a bout of pneumonia can be a life–threatening condition.

The same is true for the ocean. The more factors that are negatively impacting the ocean's health, the more susceptible and less resilient it becomes. On the flip side, the more of these issues we can identify and resolve, the stronger the ocean becomes, making it better equipped to fight the bigger, harder-to-fix issues such as ocean acidification and climate change.

The first step toward solving any problem is to identify it. In the following sections, we identify the most serious problems negatively impacting the ocean's health.

Pollution

Remember when ocean pollution made the headlines leading up to the 2016 Summer Olympics in Rio de Janeiro? Stories of raw sewage found flowing from Rio's poor neighborhoods right into the Guanabara Bay, where many of the competitions were scheduled to be held, were everywhere. Testing showed that Rio's waters contained viruses and bacteria up to 1.7 million times more hazardous than what would be considered acceptable levels in the United States or Europe. Tons of trash was also floating in the bay, and nearby industrial plants had been dumping toxic waste into the bay for decades, affecting the health of not only wildlife but also people. Unfortunately this was not an isolated instance. Pollution like this, at varying levels, flows into our ocean all over the world every day.

Pollution poses a catastrophic threat to the ocean's health, and it comes in many forms from numerous sources.

Plastic

Plastic is a petroleum-based product created by humans and used in nearly every product imaginable — grocery bags, car seats, cellphones, televisions, furniture, building materials, packaging, containers, and even some toothpastes and beauty products. In many ways, plastics have improved our lives, but they also negatively impact our health and our environment, including the ocean.

Every year about 8 million metric tons of plastic enter our ocean (though some estimates are much higher). That's the equivalent of at least one NYC garbage truck dumping a full load of plastic directly into our ocean every single minute of every day for a year. And that's on top of the estimated 100 million metric tons of plastic already floating around in our ocean (see Figure 21-2).

Except for a few litterbugs, most people don't intentionally pitch their plastic into the ocean. That would be crazy! So, where does all this plastic come from? Unless the plastic is recycled or buried in a landfill (where it lasts a very, very long time), it could be blown by the wind or carried by the water via storm drains, streams, and rivers to the sea; unfortunately, there are many sources of plastic, including the following less obvious ones:

>> *Nurdles* are plastic pellets that are melted down to make nearly all plastic products. They're small and light, making them easy to transport, but also easy to lose, difficult to clean up, and susceptible to being blown or washed into the ocean.

>> *Microbeads* are tiny pieces of plastic used in some toothpastes, face scrubs, and home cleaning products. They're so small they're really hard to filter, and they make their way back into the drinking water system or to lakes, streams, rivers, and the ocean, where they're consumed by fish and other wildlife.

>> *Synthetic fibers,* including polyester, nylon, and acrylic, are used to make about 60 percent of clothing worldwide. Each time these garments are washed, thousands to millions of these plastic fibers slough off and are sent into our water systems. These fibers also enter the environment when the clothes are discarded.

FIGURE 21-2:
Ashlan and Philippe Cousteau after slogging through a trash-littered mangrove swamp in Belize while filming a documentary.

Source: Ashlan Cousteau

"But, Ashlan and Philippe, I recycle!" We know you do, but much of the plastic used at home isn't recyclable, and even when it is, countries that used to outsource their recycling to countries like China and India can no longer do so because they're are refusing to take it (they are already so overloaded). Also, many countries have no organized waste management, let alone recycling. So where does all that plastic go? Well, it usually ends up in a landfill or in the ocean.

REMEMBER

The big problem with plastics is that they never go away; they only break down into smaller and smaller pieces through *photodegradation* (see the nearby sidebar), wind, and waves. Those tiny pieces are never truly gone (see Figure 21-3), many contain toxic chemicals such as bisphenol A (BPA) and Polystyrene oligomer, and they absorb other toxins such as PCBs and dioxins from the water. (If you don't know what these substances are, just trust us, they're very bad for anything that consumes them, including you and marine life.)

FIGURE 21-3:
Microplastics found in 1 square meter (3 square feet) of beach in Melbourne, Australia, by teacher fellows during an EarthEcho International expedition.

Source: EarthEcho International – www.earthecho.org

Microplastics comprise 85 percent of plastic found on our shorelines. Worse, they look like plankton (see Chapter 7), so they're eaten by small fish or other sea creatures, which are eaten by bigger fish, and so on. Through a process called *bioaccumulation*, the concentration of microplastics and the toxins in them, increases up the food web, eventually landing on your dinner plate.

PHOTODEGRADATION VERSUS BIODEGRADATION

Photodegradation refers to the breaking down of a material by light energy, but it usually involves a combination of light and air. Through photodegradation and mechanical energy (wind and waves), plastics are broken down into smaller and smaller pieces, but they are never broken down into the atoms or molecules from which the plastic was originally made. *Biodegradation* is breakdown by living organisms, such as bacteria, which break down the material into the chemical components from which they were made.

Even if plastics don't end up on your plate, seabirds, sea turtles, and other marine animals you know and love often eat plastic thinking it's food. A floating plastic bag or a balloon may look just like a jellyfish to a hungry turtle. These animals get plastic stuck in their throats, noses, or digestive systems, or get wrapped up in it, choking or drowning them or even causing starvation as the plastic fills their digestive tracts.

Oil spills and other oil pollution

Oil spills are devastating, especially when they occur in the ocean. The 2010 BP Deepwater Horizon oil spill released 4.9 million barrels of oil into the Gulf of Mexico (according to the U.S. government, but many researchers think the amount was much higher). It is considered the largest marine oil spill in the history of the petroleum industry to date (see Figure 21-4).

Oil is noxious and sticky. It prevents birds from flying, chokes dolphins and other marine animals, and causes hypothermia in otters. But not all the destruction is easy to see. Petroleum's harmful chemicals cause lesions, disrupt reproduction, and cause birth defects in marine life (not to mention us), thus impacting future generations.

But oil spills aren't the only source of oil pollution. Every year, hundreds of millions of gallons of oil enter the ocean from sources of consumption, such as oil leaking from cars and trucks onto roadways and washing down storm drains and eventually to the ocean, and oil leaking from airplanes, small boats, jet skis, and other watercraft.

Sunscreen

According to the National Park Service, 4,000 to 6,000 tons of sunscreen enter reef areas annually. That's a lot of SPF, and many of the active ingredients in some sunscreens are very harmful to coral reefs. The worst offenders are chemicals such as benzophenone-3 (oxybenzone) and octinoxate, which are found in many SPF products. They're also absorbed by the skin and enter the bloodstream easily, posing a possible health threat to humans.

FIGURE 21-4:
Philippe
Cousteau
kneeling on a
beach covered
in oil from
the BP
Deepwater
Horizon
oil spill.

Source: Philippe Cousteau

REMEMBER

Look for mineral sunscreens containing titanium oxide or zinc oxide as their active ingredient. These are nontoxic and also don't get absorbed into the skin, which is better for you, too. However, avoid products with nano titanium or nano zinc, because they *can* be absorbed through your skin.

Runoff

Runoff is rainwater, melted snow, or irrigation water that's not absorbed into soil and instead runs into the local watershed: streams, rivers, ponds, lakes, aquafers (all of which contribute to our drinking water) and then often into the ocean taking with it pollutants and chemicals. Runoff can be broken down by source:

>> **Agricultural (farm) runoff** may contain high levels of animal waste, fertilizer, pesticides, heavy metals, nitrogen, phosphorus, antibiotics, and other potentially harmful substances. Pesticides can build up in fish and ultimately end up in the diets of people and pets. Polluted agricultural runoff can also

trigger *harmful algae blooms* (HABs) in coastal waters, which pose a health risk to humans, livestock, and wildlife. (See Chapter 8 for more about HABs.)

» **Urban (stormwater) runoff** carries oil, grease, pesticides, herbicides, trash, pet waste, and other toxic or potentially harmful substances from pavement (roads, sidewalks, driveways, parking lots), rooftops, and construction sites down storm drains and into lakes, streams, and rivers, and eventually the ocean. Water flowing through storm drains is mostly untreated, and if it does flow through a treatment facility, it can overload the facility and carry human waste into the sea as well.

» **Wastewater (sewage)** is all the used water that flows out of homes, apartments, public restrooms, hotels, motels, restaurants, schools, hospitals, laundromats, car washes, and so on. Depending on the source, it may contain feces, urine, detergents, household or industrial chemicals, and anything else flushed down a toilet or poured down a drain. Some municipalities thoroughly treat wastewater and then put that clean water safely back into the public waterworks system. Some homes have their own treatment facility (a septic system) that filters out most of the waste and returns the cleaned water to the ground. Some municipalities treat the water to varying degrees and send it out into natural water sources (lakes, rivers, ocean).

In some areas, untreated waste water is released directly into nature, usually the ocean. This is common practice for some small islands, less developed countries, and even some areas in developed countries. The contaminated water leads to fish die-offs, dead zones, and HABs. Trust us, you don't want to swim in *that*. Lobsters, on the other hand, love to forage around sewage outflow pipes. Think about that next time you see lobster tail on the menu!

Sound pollution

The ocean has often been referred to as the silent world, but this peaceful place has become ear-splittingly loud. Ships, oil drills, sonar devices, and seismic tests have made the sea loud and chaotic. Sound waves can carry underwater undiminished for miles. Whales and dolphins are particularly impacted by noise pollution. These marine mammals rely on *echolocation* to communicate, navigate, feed, and find mates, and excess noise interferes with their ability to do all of that.

Cetaceans (certain marine mammals) aren't the only ones that suffer from noise pollution. Other marine animals such as clown fish use sound to navigate, find food, attract mates, and avoid predators. A noisy ocean is a less healthy ocean for all of us.

Light pollution

Light pollution on land can affect marine animals. For example, when baby sea turtles hatch at night, they crawl toward the brightest light, which should be the moon and its reflection on the sea. But with coastal development, many babies are attracted to the bright lights of homes and businesses and end up crawling in the wrong direction and dying. (If you're in a sea turtle nesting area, please either turn off your lights or use red lights at night.)

Artificial light also penetrates the water's surface creating a vastly different world for sea creatures living in shallow reefs near urban areas. Light disrupts the normal cues associated with *circadian rhythms* (sleep-wake cycles associated with the movement of the sun and moon) and the timing of migration, reproduction, and feeding. Furthermore, artificial light at night from urban areas, boats, oil rigs, and so on can make it easier for predators to find small fish to prey on, and negatively impact breeding in reef fish.

Overfishing

According to the Food and Agriculture Organization (FAO) of the United States, 90 percent of fish stocks (the highest percentage ever recorded) are now either fully fished or overfished at biologically unsustainable levels (see Figure 21-5). The stress on fish populations has led to several problems, including the following:

» Limited availability of an important source of protein for more than one billion people around the world.

» *Illegal, unreported, and unregulated* (IUU) fishing, which fleeces the economy and environment, impoverishes coastal communities, and has sinister links to drugs, human trafficking, terrorism and even slave labor. Illegal fishing accounts for about 20 percent of the total fish caught.

  The number of people put to work on fishing boats against their will is unclear, but according to the International Labour Organization, they account for a significant number of the 21 million people around the world who are trapped and enslaved.

» Fish fraud. Approximately 30 percent of seafood globally is mislabeled on purpose (a practice referred to as *species substitution*), primarily because many popular seafoods are overfished and unavailable. This fraud cheats consumers out of what they paid for and puts public health and the oceans at risk. Fish fraud also allows for illegally caught fish and fish caught using slave labor to be laundered into the legal seafood trade.

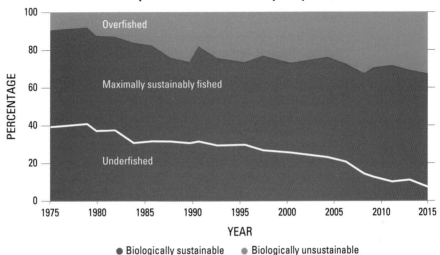

World aquaculture of food fish and aquatic plants, 1990-2016

Overfished

Maximally sustainably fished

Underfished

● Biologically sustainable   ● Biologically unsustainable

Source: Food and Agriculture Organization of the United Nations. Reproduced with permission.

FIGURE 21-5:
Graph of global fisheries' decline.

Acidification

The ocean is a giant carbon sink. It has absorbed about 25 percent of the excess carbon we have pumped into the atmosphere. That's good, right? It helps prevent global warming? Well, yes and no. Extracting carbon from the atmosphere is beneficial for mitigating global warming, but the carbon dioxide removed from the atmosphere interacts with water molecules to create carbonic acid, which then reacts with carbonate ions to form bicarbonate, reducing the amount of carbonate in the ocean, which is bad. Why does this matter? The problem is that animals (coral, oysters, lobsters, and even some plankton) which need to bind carbonate ions with calcium to build their shells are left without enough carbonate ions to do it.

REMEMBER

Think of ocean acidification as osteoporosis of the sea. As the ocean becomes more acidic (less alkaline), shells of marine organisms that are made from calcium carbonate become more brittle and fragile, making the organisms' survival more difficult. The ocean needs to maintain a stable, slightly basic (alkaline) pH level in order for organisms that make their shells from calcium carbonate to survive.

Already, oysters in the Pacific Northwest are dying out, threatening a multimillion-dollar industry. Scientists have watched with horror as in certain regions of the North Pacific, *pteropods*, free-swimming ocean snails crucial to the ocean food chain, are literally dissolved before their eyes due to the higher acidity in the ocean. Ocean acidification is altering the very basis of our entire marine food web. That is terrifying for all of us. (Visit oceanacidification.noaa.gov/Home.aspx for additional information about ocean acidification.)

Habitat destruction

Like people, marine organisms need a place to live and raise their families — a habitat. Natural events and human activities can destroy these habitats, negatively impacting the organisms' ability to survive and multiply. Habitat destruction occurs when wetlands are dredged and drained for residential or commercial development; inland dams block fish migration routes and curb fresh water flow, thereby increasing coastal salinity; and deforestation causes erosion, which sends silt downstream and onto coral reefs, blocking their exposure to sunlight.

Destructive fishing practices also destroy habitats; for example, *bottom trawling* involves a ship dragging a giant weighted net along the seafloor pulling up everything in its wake — and we mean EVERYTHING. Honestly, it's so horrible, it should be outlawed (see Figure 21-6 for a comparison). *Dynamite fishing* (yep, tossing dynamite into the water where the explosion stuns the fish and they float to the surface) and *poison fishing* (poisoning the water, often with cyanide, to either stun the fish or kill them without leaving any marks) also obviously destroy habitats in the process.

Invasive species

Ecosystems (communities of plants, animals, and other life-forms coexisting in a given environment) take considerable time to develop a delicate balance. When a stranger from another ecosystem (an invasive species) is introduced, it can wreak havoc on that delicate balance. Lionfish are a perfect example (see Figure 21-7). Originating in the Indo-Pacific, they evolved in an ecosystem that kept their population in check. Their prey adapted in ways to avoid them, and predators adapted to eat them, despite their venomous spines. However, since they've been introduced to the Caribbean, Atlantic, and Gulf of Mexico, they've been decimating populations of native fish. Lionfish are voracious eaters, incredible reproducers, and have no natural predators in those waters.

Invasive species can hitch rides on ship hulls, in ballast tanks, through the seafood trade, and even through the aquarium trade. Once in a new place, invasive species can take over, beating out the locals for space, sunlight, and food, or simply eating them all up!

Warming sea temperatures

As we explain in Chapter 2, the atmosphere affects our ocean, and the ocean influences our atmosphere. Just as the ocean is a carbon sink (see the earlier section "Acidification"), it's also a heat sink. Water absorbs and retains heat. You can observe this quality of water when you're bringing a pot of it to boil on your stove. Even with the burner cranked to High, the water absorbs the heat for several minutes before it starts to boil.

(a)

(a) Photo by Mareano - Havforskningsinstituttet. Licensed under CC BY-SA 4.0
(b) Photo by Mareano - Havforskningsinstituttet. Licensed under CC BY-SA 4.0

FIGURE 21-6:
A healthy deep-sea ecosystem before trawling (a) versus after trawling (b).

(b)

FIGURE 21-7:
A lionfish.

Source: Philippe Cousteau

As Earth's atmosphere warms, so does its ocean. The seas have been warming at a rate of 0.2 degrees Fahrenheit every ten years, which may not seem like a lot, but it is. Increasing temps affect weather patterns, result in stronger storms, and negatively impact marine life, especially coral. Unlike humans, who have the ability to regulate their body temperature, most marine animals have a narrow temperature range they can tolerate. And because they don't have air-conditioners and fans to cool them off when it gets too hot, they can't escape the heat. Coral reefs are a great example. When ocean temperatures get too hot, coral can't survive because they lose their symbiotic zooxanthellae algae, and "bleach" (see Figure 21-8). Coral bleaching due to warming oceans is devastating coral reefs around the world. See Chapter 5 for more info on coral bleaching.

FIGURE 21-8:
A coral reef before and after a bleaching event. Notice that after a sustained period of bleaching, the coral dies and is covered by algae.

Source: Bleached Reefs in American Samoa © Underwater Earth / XL Catlin Seaview Survey – www.underwater.earth

Also, remember what you learned in science class? Generally, substances expand with heat, and the same is true of water (though water is somewhat unique in that it expands when it freezes, too, which is why ice floats in water). Ocean warming is one of the main causes of increasing sea levels, which creates all sorts of problems, including habitat destruction for humans.

Coming Up with Solutions

Environmental issues impacting our ocean can be overwhelming to the point where it's easy to be frozen into inaction. Fortunately, with a healthy combination of increased awareness, human ingenuity, and nature's resilience, humans can collaborate with nature to solve the ocean's problems, restore its health, and preserve it for future generations. In this section, we introduce possible solutions to the problems described in the first part of this chapter, and what you can do to help.

Improving socioeconomic conditions

As the Ocean Health Index (OHI) reveals, areas of the world with stable and effective governance score much higher in terms of sustainable ocean health than regions that suffer from chronic corruption, dictatorship, civil strife, war, and poverty. In stable countries with thriving economies, communities have the financial resources to build and maintain modern sewer systems, water treatment plants, and waste-management infrastructure; they designate parcels of land to preserve nature; they have regulations and enforcement in place to govern the use of the ocean's resources; and they allocate resources for ocean conservation.

In impoverished and unstable areas, people are often just trying to survive. It's not necessarily that they don't appreciate the ocean, rather that they often don't have the capacity and financial/societal/governmental infrastructure to be able to implement effective environmental protection. When you're worried about feeding your children today, it's hard to plan for tomorrow. Tragically, this often means that people are living in toxic environments themselves. As OHI points out, "Improving ocean health will require efforts from all sectors to promote peace, justice, gender equality, socially responsible business, and other aspects of civil health, because progress in those areas makes it much easier for communities and nations to improve the environmental and economic conditions needed to boost ocean health." This notion of environmental justice is not exclusive to developing countries. Here in the United States, for example, people who live, work, and play in America's most polluted environments are commonly people of color and/or the poor.

Building marine sanctuaries

As of the publication of this book, only 6 percent of the ocean is meaningfully protected as a marine reserve or *marine protected area* (MPA) — a place where no destructive or extractive activities such as fishing or mining can take place. Marine reserves cannot protect against the effects of climate change (such as ocean warming and acidification), but they can help to rebuild species abundance, diversity, and restore marine life by eliminating overfishing and habitat destruction. These healthier, more vibrant areas then have a better chance at fighting the effects of climate change and ocean acidification. Marine sanctuaries are the best tools available for restoring ocean health. We have witnessed the incredible resilience of nature with our own eyes, as told in the nearby sidebars.

MPAs deliver numerous benefits to ocean environments and marine life, including the following benefits highlighted by the organization Ocean Unite (www. oceanunite.org):

>> More fish, bigger fish, and more types of fish both within the reserve and also outside due to the "spill-over" effect.

>> The recovery of areas that have been damaged and an increase in marine life (not just fish) within the area.

THE BIKINI ATOLL BOUNCES BACK

In 2016, we traveled to the Marshall Islands, home of Bikini Atoll — a primary nuclear weapons testing site for the United States during the Cold War. For 12 years starting in 1946, 23 nuclear bombs were detonated inside or over the atoll and underwater. The largest bomb, Castle Bravo burned so hot that the sand turned to glass. It was one thousand times more powerful that the bomb dropped on Hiroshima. As you can imagine, these bombs wiped out every single living thing for miles around.

Sixty years later, we went to Bikini Atoll to see how nature and more specifically its shark population were doing. What we found was remarkable. Not only had nature recovered, but it was flourishing. Giant clams, grouper, coral, and fish were everywhere. And the grey reef shark population was thriving. At one point, we were swimming with 70 sharks surrounding us. And to think that just six decades ago we incinerated that atoll, wiping out all life on it and around it.

Because the above-water part of Bikini Atoll remains radioactive no one goes there, so it has become a de facto marine reserve. Without human interference, the tiny island was able to bounce back from dead zone to Pacific paradise in just 60 years.

>> The prevention of coastal erosion and mitigation of the impacts of natural disasters (such as hurricanes) through the protection of coastal habitats such as mangroves and coral reefs.

>> The sequestration and storage of carbon also through the protection of coastal habitats such as mangroves, seagrass beds, and salt marshes.

>> Reduction of poverty through providing food and employment for some of the billions of people around the world who directly rely on a healthy ocean for survival.

>> High financial returns. Protecting 30 percent of the ocean has been estimated to cost between US$223 and $228 billion, but it has been estimated that the financial net benefits from the increased ecosystem goods and services (once all costs have been taken into account) will range from US$490 to $920 billion by 2050. In the financial world, that's a return on investment no one would turn down!

>> A global review of the impacts of marine protected areas on fish found that fish biomass (weight) increased by 446 percent, it was denser (more fish) by 166 percent, species size increased by 26 percent, and there were 21 percent more types of fish. Source: Ocean Unite 2020; https://www.oceanunite.org/issues/marine-reserves/.

THE POWER OF ONE

About an hour and a half north of Cabo San Lucas, at the bottom of the Baja Peninsula, is the tiny town of Cabo Pulmo. In the 1970's the patriarch of this small community, a man named Juan Castro, started to get worried. The fishing that had sustained his community for generations was declining. Every day they had to venture farther offshore to find fish with no guarantee of success. Like any good father, Juan worried for his ability to feed his family and for the future of his children. Then, one hot summer day, a group of tourists drove up from Cabo offering him money to take them out on his boat. Much to his surprise, they did not want to fish; instead, they had come to scuba dive. When they surfaced, one of the divers passed his mask to Juan and invited him to look underwater. This was a pivotal moment and changed Juan forever because, despite growing up next to the ocean, Juan had never looked beneath the surface.

The beauty he witnessed gave him an idea, and after years of work, he convinced the local community to join him to shift their local economy to tourism, which turned into a crusade to create a marine protected area in their local waters that would ensure a healthy reef for tourists to enjoy.

(continued)

(continued)

Fast-forward more than 20 years, and the 70-square-kilometer no-take marine reserve in Cabo Pulmo is a paradise unlike anything else in the Sea of Cortez. With upwards of a 1,000-percent increase in biomass (volume of living creatures) in the area, even for people like us who have been diving all over the world, it's magical. When you talk to Juan and his children and grandchildren, you hear a common refrain — they're proud of what they've achieved and are most proud because, in the words of one of his sons, Mario, they have something they can pass on to their children that will only increase in value.

REMEMBER

One particular global movement that encourages the establishment of more MPAs is 30x30, which is based on the calculations by the preeminent Harvard biologist, E.O. Wilson. Wilson said that to stave off the catastrophic global decline in biodiversity biodiversity, we must protect 30 percent of the planet by 2030. For more info about the 30x30 movement, visit www.oceanunite.org/30-x-30.

Cutting greenhouse gas (GHG) emissions

Greenhouse gasses produced by human activities, including carbon dioxide, methane, nitrous oxide, ozone, and CFCs, harm the ocean by making it warmer. Carbon dioxide harms the ocean in a second way — by increasing the acidity of seawater (there's ocean acidification again). To combat this, we need to reduce GHG emissions, especially carbon dioxide. To make this task more manageable, we can break it down into three areas: where we live and work, how we travel, and what we eat.

>> **Where we live and work:** Fossil fuel use in buildings represents about 28 percent of all U.S. building energy emissions. To bring that number down and save money in the process, practice energy efficiency — choose energy-efficient lighting and appliances, seal and insulate your home and office, and program your thermostat, to name just a few techniques. When possible, opt for a high-efficiency electric furnace or water heater over one that burns fossil fuels. Look for ways to implement the use of renewable energy sources, such as solar, wind, hydroelectric, biogas, and geothermal.

>> **How we travel:** GHG emissions from transportation make up about 29 percent of all GHG emissions in the United States, according to the EPA. This makes transportation the largest contributor of GHG in the U.S. Planes, cars, buses, trains, and shipping all add up. Airlines, like United, are starting to use sustainable aviation biofuels in their planes; electric cars and charging stations are now widely available (and fast!); and cities are working hard to update and improve their mass transport systems. When you need to travel, opt for the greenest option available. Better yet, consider whether you even need to travel. One of the outcomes from the coronavirus is that it showed just how many of us can work productively at home (and avoid the daily commute to work and back).

>> **What we eat:** Every food has a carbon footprint, but those footprints differ dramatically. The World Bank and International Finance Corporation (IFC) found that livestock agriculture accounts for nearly half of all human GHG emissions. Eating an 8-ounce steak is the equivalent of driving a gas-powered car for 14 miles. Beef, pork, lamb, and cheese have high carbon footprints compared to fruits, vegetables, beans, and grains. The best way to eat for the ocean and planet is vegan or vegetarian, but even eating vegetarian or vegan one meal a week (or once a day) can make a big difference. Philippe and I are mostly vegetarians, but we have two pet chickens who give us incredible eggs. We also eat cheese and splurge on local, humane meat about twice a month.

Restoring and conserving coastal and ocean habitats

Restoring and conserving coastal habitats is a win for the ocean, marine life, and humans, too. Just consider the following facts:

>> Healthy coastlines produce at least as much food per acre as farmland.

>> Tidal marshes, wetlands, seagrass beds, sand dunes, and mangrove forests protect the coastline against damage from wind, waves, and flooding.

>> Healthy estuaries and coastal habitats help to prevent outbreaks from toxic microbes that cause massive fish kills and illness in anyone coming into contact with the water.

REMEMBER

Blue carbon is the carbon captured by coastal ecosystems, mostly by mangroves, tidal (or salt) marshes, and seagrasses. When these systems are healthy, they pull carbon out of the atmosphere. When they're destroyed, not only do they stop extracting carbon from the atmosphere, but the carbon they've stored for thousands of years can be released into the atmosphere. Even though these ecosystems only account for 2 percent of our ocean habitat, they're incredibly important. Visit www.thebluecarboninitiative.org for more info. Another new concept of blue carbon is the idea that restoring large marine wildlife such as whales and sharks would help to decarbonize the atmosphere by recarbonizing the biosphere. See Chapter 19 for more.

>> Every million dollars spent on coastline restoration and preservation generates 17 jobs as contrasted to the 5.2 jobs on average generated by each million dollars invested in oil and gas development.

Reducing the impact of plastics and other trash

With about 100 million metric tons of plastics floating around in our ocean and even more on its way, this problem requires a multipronged solution. Such a solution must include reducing the use of plastic products, especially single-use products, such as plastic bags, water bottles, plates, cutlery, packaging, and straws; improving waste collection, infrastructure, and management; and expanding recycling, particularly in the countries where most of the plastic originates.

Five countries in Asia — China, Indonesia, the Philippines, Thailand, and Vietnam — account for as much as 60 percent of the estimated plastic waste entering the ocean that we're aware of. But *every* country in the world can do better, including the United States. We need to establish a circular, not a linear, economy when it comes to plastic. That means creating an economy where resources, such as plastics, are used, recovered, and reused over and over again, instead of heading directly to a landfill or the sea.

REMEMBER

The best way to cut back the plastic in the ocean is to stop making and using so much of it in the first place. Most of all, you can just say no to single-use plastics, which we use for only a few short minutes but last forever. In the United States, one important step you can take is to contact members of the U.S. Congress and encourage them to support the Break Free From Plastic Pollution Act (visit oceanconservancy.org/action-center for details).

Preventing and recovering from overfishing

Overfishing is one of the biggest threats to ocean wildlife abundance and diversity, but it may be one of the easiest problems to solve because viable solutions exist if people would only put them into practice:

» Smart, practical fish policies, such as the Magnuson-Stevens Fishery Conservation and Management Act in the United States, designed to prevent overfishing, rebuild overfished stocks, increase long-term economic and social benefits, and ensure a safe and sustainable supply of seafood. Visit `www.fisheries.noaa.gov/topic/laws-policies` for details.

» Stopping subsidies to industrial fishing operations. Fishing subsidies are estimated to be as high as US$35 billion worldwide, $20 billion of which directly contributes to overfishing of an already depleted resource.

» Prohibiting inefficient or destructive fishing practices, such as trawling, and requiring the use of techniques and devices to limit *bycatch* (unwanted fish and other marine animals caught while fishing for other species).

» Implementing traceability standards to prevent fish caught illegally from being imported.

» Consumer education and technology, such as the Monterey Bay Aquarium *Seafood Watch Guide* (`www.seafoodwatch.org`), to enable consumers to make well-informed seafood choices; for example, not ordering fish on the menu that are overfished.

» Establishment of marine sanctuaries (no-fishing zones) to enable restoration of fish populations.

Engaging youth

This country, and the world at large, needs a wholesale cultural and social revolution to achieve what needs to be done to save our oceans for future generations. The best way to achieve that is through something the environmental movement has neglected for far too long — education. The only way to build lasting change is from the ground up. Youth are the only ones who have the ability to influence not only their own behavior and the behavior of their peers, but also the minds and hearts of parents, teachers, government representatives, community leaders, and business leaders.

The key to effective education is realizing that young people aren't just the hands and feet of the environmental movement, to be told what to do by adults. They must be empowered to be the hearts and minds, to be leading us, because when they do, remarkable things can happen. But for education to be effective it must be inclusive which means engaging youth in communities that have traditionally been left out of this conversation.

EarthEcho International (`www.earthecho.org`), is a leading nonprofit dedicated to building a global youth movement to protect and restore our ocean planet. I (Philippe) founded EarthEcho in honor of my father Philippe Sr. and to date more than 2 million people in 146 countries have participated in our programs. From water quality, to ocean health, to enhancing biodiversity, our programs provide original content, immersive experiences and trusted resources that equip youth and educators to take action in their communities and around the world to build the foundation of a thriving future. Our youth leaders have passed laws, raised critical funds, started movements to protect land, founded successful businesses that help people and the planet, and so much more (see Figure 21-9).

FIGURE 21-9: Ashlan and Philippe learn from youth leading an oyster restoration project in the Chesapeake Bay during an EarthEcho International expedition.

Source: EarthEcho International - www.earthecho.org

EarthEcho works with youth all over the world who have the optimism and determination to make the future better and more just. They recognize that when we come together there's nothing we can't achieve, no problem we can't overcome, and no one who can stop us from building a better world. More than anything else, our work with youth is what gives us hope for our ocean.

Get involved!

Volunteer at a nonprofit like EarthEcho International and countless others across the globe (see Chapter 24 for a list of some of our favorites). No matter where you live in the world or how far away you are from the ocean, your actions matter. Cleaning up your local watershed, helping replant trees, raising money, support conservation politics and so much more — it all ultimately helps our sea.

REMEMBER

Most important: VOTE. Engage in the democratic process and encourage local and national leaders to deliver policies that promote responsible stewardship of our amazing ocean resource. As they say, Vote the Ocean.

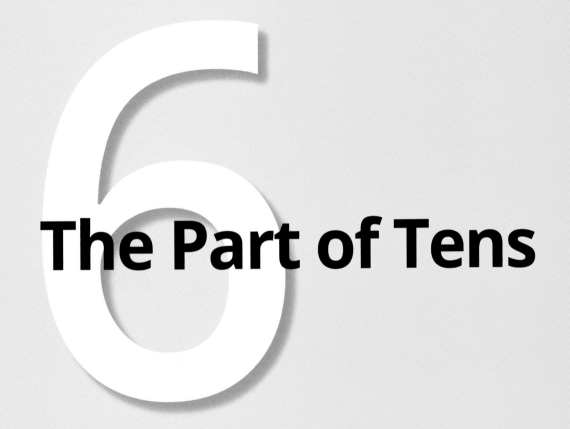

The Part of Tens

IN THIS CHAPTER

» **Recognizing the two reptiles you need to watch out for**

» **Knowing the one whale to avoid**

» **Distinguishing dangerous from harmless sharks**

» **Naming two fish that can really put the hurt on you**

» **Identifying the ocean's most notorious species**

Chapter **22**

Ten Deadly Ocean Creatures

The ocean is our favorite hangout, colorfully decorated and home to some of the most fascinating characters, some of which are potentially dangerous. To stay safe, you need to watch your step, keep your hands to yourself, respect the residents, and steer clear of some of the more dangerous denizens. In this chapter, we introduce you to nine sea creatures (in no particular order) to appreciate from a distance and then the number one most deadly creature on land or sea.

REMEMBER

Our goal in this chapter is to inspire awe and respect, not fear. Every creature mentioned in this chapter has adapted to survive and thrive in its environment. Some are large and fast with powerful jaws and big teeth, others use camouflage to their advantage, some are really smart, some carry toxins, and others pack a weapon, but we love them all and hope you do, too.

Note: We want to give a special shout out to our dear friend and a person with a *super* cool job, Prof. Jamie Seymour, PhD, at James Cook University. As a leading toxicologist/venomologist, he works with many of these deadly animals every day. Talk about brave.

Saltwater Crocodiles

Saltwater crocs can grow in excess of 6 meters (20 feet) and about 900 kilograms (2,000 pounds). They have big teeth, powerful jaws, can run about 24 kilometers per hour (15 mph) in short sprints and swim about 32 kilometers per hour (20 mph) You may be able to outrun one, but you can't outswim one.

Three characteristics make them particularly deadly. First, they're aggressive. You don't need to poke 'em with a stick to make 'em mad. All you need to do is invade their personal space. Second, they ambush their prey. They hang out very still at the water's edge, hiding just below the surface of the water, or hidden in the mud until an unsuspecting monkey, boar, water buffalo, or other animal drops by for a drink — then bam! They clamp down and won't let go. Third, immediately after clamping down, they do a death roll to drown and dismember the body, so if the initial attack doesn't kill you, you'll soon wish it had.

Saltwater crocs have even been known to attack and eat sharks. We're guessing the winner would come down to size and to who surprised whom. (See Chapter 13 for more about saltwater crocs.)

Fugu Fish

Fugu fish, a type of pufferfish or blowfish, are considered by some to be a delicacy. However, if not prepared properly, this fish could be the last one you ever eat. When threatened, this species of pufferfish (like many others) can blow itself up to several times its normal size. If the predator doesn't take the hint and ends up eating the pufferfish, it's in for a rude surprise. This particular species of puffer-fish has a toxin in its liver that makes it taste horrible and is potentially deadly. Fugu is considered a dangerous delicacy because, if prepared carelessly, the parts of the body that carry the toxin can end up in the meal. In fact, this cute little puff of joy carries enough toxin to kill up to 30 adult humans, and it has no known antidote.

Eating fugu is like playing Russian Roulette with your dinner. Tetrodotoxin, stored in the liver, ovaries, and other parts of this fish, may be more toxic than cyanide. A small dose of tetrodotoxin could result in numbness in your mouth, vomiting, paralysis, and possibly death. We recommend crossing it off your menu.

The pufferfish is poisonous, not venomous. What's the difference? A poisonous animal is toxic but doesn't actively deliver the toxin — you have to touch it or eat it. In contrast, venom is delivered intentionally; for example, through a bite or sting. Some animals, such as a spitting cobra, are *toxungenous*, meaning they spit, spray, or fling their toxins.

Killer Whales

Their name says it all. Killer whales (orcas) are one of the few mammals that kill for fun. In fact, they hunt in packs and teach their calves how to hunt from an early age. They're also big, powerful, and smart, making them one of the most formidable ocean predators. In a meet-up between a killer whale and a great white shark, the killer whale would likely prevail. They've even developed a special technique for dealing with sharks — biting the shark and turning it on its back to put the shark in a coma-like state called *tonic immobility*. Told you they were smart. (See Chapter 15 for more about killer whales.)

While orcas in captivity have fatally wounded people (another reason to #Empty-TheTanks), none has ever been reported to have killed a human in the wild.

Blue-Ringed Octopus

Beautiful and deadly, the blue-ringed octopus is named after the vibrant blue circles it shows when feeling threatened (see Figure 22-1). Though docile, their bite contains the powerful neurotoxin called tetrodotoxin (the same toxin carried by the fugu pufferfish). The octopus uses the venom to immobilize its prey for an easier meal that doesn't wiggle around. Fortunately for the octopus eating this tainted meal, it's immune to its own venom, but you most certainly are *not!* (See Chapter 10 for more about octopi.)

FIGURE 22-1:
Blue-ringed
octopus.

Source: Sheree Marris – www.shereemarris.com

Sea Snake

While snake enthusiasts can't agree on the most venomous snake, they do agree that sea snakes are super venomous. But venomous and deadly are two different things. Sea snakes are very docile, and they don't have huge front fangs like those of cobras, so they're not likely to bite you. Even if they do, they may not deliver enough venom to kill you. However, if you were to get a good dose of it, good luck. It's highly concentrated. (See Chapter 13 for more about sea snakes.)

Stone Fish

The stone fish holds the title for most venomous fish in the world, but you may not even notice them, because they look like rocks (see Figure 22-2). At least not until you step on one. Then, holy fugu fish, do they ever pack a lethal punch! The first sign of trouble is that your foot becomes impaled on the spikes of their dorsal fins (exceedingly painful). Then, the toxin enters the bloodstream, causing swelling around the wound, difficulty breathing, irregular or no heartbeat, nausea, vomiting, delirium, and even death.

FIGURE 22-2:
A stone fish.

Source: Prof Jamie Seymour – James Cook University

The good news is that as long as you don't step on one, or accidently grab it, you have little to fear.

Sharks (But Not All of Them)

Many sharks are apex predators that play a key role in maintaining a healthy ocean, as we explain in Chapter 12. These predators are big, strong, and fast and are equipped with big, sharp teeth and strong jaws. But some sharks pose a greater threat than others. Some of the deadliest (to humans) sharks include the following:

» Great white shark (see Chapter 12)

» Tiger shark (see Figure 22-3)

» Bull shark

» Oceanic whitetip shark

FIGURE 22-3:
A tiger shark.

Source: Michael Muller – www.mullerphoto.com

However, the vast majority of sharks are relatively harmless to humans, including the following:

>> Whale shark

>> Nurse shark (as long as it's not harassed)

>> Basking shark

>> Leopard shark

>> Angel shark

>> Bamboo shark

REMEMBER

With only a handful of fatal shark bites each year, even predatory sharks pose a negligible risk to humans. See Chapter 12 for the list of things that are much more likely to kill you than a shark.

Cone Snails

Yep, this beautiful little snail is a vicious predator. Experts hypothesize that a single cone snail contains enough venom to kill 700 adult humans. Like other snails, the cone snail is slow, so it uses a harpoon-like tooth to inject venom into its prey, quickly immobilizing the prey so the snail can eat it. If you're on the receiving end of one of the snail's venom-tipped harpoons, head to the hospital; death can occur within one to five hours in severe cases.

While there is no anti-venom, on the bright side, scientists have used the venom to create better, non-addictive painkillers and insulin. See Chapter 10 for more on Cone snails.

Box Jellyfish

While people tend to fear sharks the most, if they're going to be afraid of anything in the ocean, it should be the box jellyfish (see Figure 22-4). They're translucent, so you can barely see them in the water. They're much better swimmers than most jellies. They have lots of eyes (24 to be precise). They deliver their venom on contact, and a sting from the most venomous of the group can result in paralysis, cardiac arrest, or even death in a matter of two to five minutes. Yikes.

Humans

Even with all these deadly creatures in the sea, from tiny cone snails to giant predators much larger than people, humans are the most dangerous creature out there. Over the past 40 years, thanks to humans, 50 percent of the

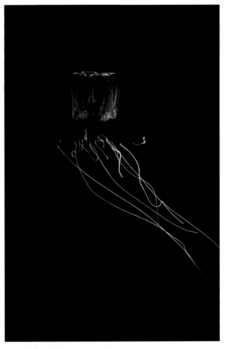

Source: Sheree Marris – www.shereemarris.com

FIGURE 22-4:
A box jellyfish.

biodiversity on Earth has disappeared, and over 1,000 species go extinct each year. Every year, humans are responsible for killing hundreds of millions of sharks; billions of pounds of fishing bycatch; and more than 300,000 small whales, dolphins, and porpoises (due to fishing line entanglement), not to mention the fact that every year humans murder about 450,000 other humans. Mon Dieu, we really are the deadliest species on the planet.

IN THIS CHAPTER

» Getting at the truth behind garbage patches and rising sea levels

» Knowing for sure whether sharks must swim continuously to survive

» Understanding what makes seawater special and whether it kills bacteria

» Discovering the truth about waterspouts and jellies

» Recognizing what makes the ocean blue

Chapter **23**

Ten Ocean Myths Busted

Any large, complex topic, such as the ocean, breeds myths, misinformation, and false assumptions. In this age of the internet and social media, people are becoming more informed about the ocean, but they're also becoming more misinformed. Sometimes the misinformation is all in good fun; for example, in 2008, as a hoax, the BBC posted a video showing a flock of flying penguins. In other cases, misinformation can be fatal; for example, the belief that because the ocean is made up of salt water, it kills bacteria. And in many cases, the misinformation is neither humorous nor harmful — it's just wrong.

In this chapter, we present ten popular myths about the ocean and reveal the truth behind each of them.

Melting Sea Ice Increases Sea Levels

Sea levels are rising, but the problem isn't due to melting icebergs. If you're extremely bored on a summer day, fill a tall glass half full with water, add four or five ice cubes, take note of where the water line is, and then watch the ice melt.

The water level won't rise. Why? Because ice displaces the same amount of water when it's frozen as it does when it's melted. This is the same in the ocean. As sea ice melts it doesn't raise the level of the ocean because it displaces the same amount of water melted as it did when frozen.

The confusion probably comes from the distinction between ice on land and ice in water. When *glaciers* (large sheets of ice and snowpack on land) melt, the resulting water flowing into the ocean *does* increase the level of water in the ocean. Ice and snow on land are always melting, but under normal (recent historical) temperatures, the melting was offset by evaporation from the ocean falling as snow and ice over the polar regions, maintaining equilibrium. However, as global temperature rises, glaciers are melting at a faster pace than they're being restored, and that extra water is flowing right into the ocean. To compound the problem, the warm water melting on the surface trickles down into cracks and fissures in the ice, causing additional melting. And global sea level rise.

Sharks Must Swim Constantly to Survive

The idea that sharks must swim constantly to survive can be broken down into two parts. First, it's true that sharks must swim to keep from sinking. Unlike bony fish, sharks, rays, and skates have no swim bladder. They do, however, have an oily liver which makes them close to neutrally buoyant, but they will still sink, albeit slowly. So, they have to keep on truckin' to keep their position in the water column, though some sharks do rest on the bottom.

Second, sharks can't open and close their gills to flush oxygenated water over them (known as buccal ventilation), as bony fish do, so people often think that sharks must swim continuously to breathe. That's true for many sharks but not all. Depending on the species, a shark may use ram ventilation, buccal ventilation, or a combination of the two:

>> **Ram ventilation:** The shark swims with its mouth open forcing water over its gills. Great whites, makos, whale sharks, and other sharks that are constantly on the move are all ram ventilators (who knew?).

>> **Buccal ventilation:** The shark uses its buccal (mouth) muscles to pump water in and out over the gills. They can close their gill slits and open their mouth, creating a vacuum that sucks in water. Then they close their mouth, constrict their throat muscles, and open their gill slits, forcing the water out for them to breathe. Nurse and bullhead sharks, and other sharks that aren't active swimmers, use buccal ventilation. Lucky them — they can breathe while snoozing.

Some sharks, including the tiger shark, can take advantage of either method.

Some Penguins Can Fly

We're going to blame this myth on the BBC. In 2008, in celebration of World Penguin Day (and the day before April Fools' Day), they released a hoax video about a recently discovered colony of penguins that can fly. According to the video, the penguins "fly thousands of miles to the rainforests of South America where they spend the winter basking in the tropical sun," which they do not. Check it out on YouTube.

REMEMBER

So, to clear the air (ha ha), penguins can't fly . . . not in the air, anyway. That's right, they *are* birds and they *do* have wings, so in a sense they do fly, but only underwater. Water is 900 times denser than air so their wings have evolved strong muscles to deal with that density and propel them at speeds up to 20 miles per hour through the water. They are a true joy to watch underwater as they look like torpedoes in tuxedoes.

Salt Water Kills Bacteria

Some people with cuts or scrapes wade into the ocean with the mistaken notion that ocean water will clean and sterilize their wounds, and some of these people don't live to tell their stories. If you don't believe that bacteria can live in the ocean, just turn to Chapter 7 to find out about microbes that actually thrive in ocean environments.

To be fair, it is true that some bacteria can't tolerate even low levels of salt, but many bacteria can. In fact, some strains of Staphylococcus (which cause potentially deadly staph infections) can survive in water that's 20 percent salt! At a 3.5 percent concentration, ocean water poses no challenge.

As for bacteria that normally live in the ocean, they can be dangerous too. One in particular called *Vibrio vulnificus*, which lives in warm salt water and is related to cholera, kills several people every year when they swim in the ocean with an open wound or cut on their body.

REMEMBER

We're not trying to scare you off the beach, but if you have a deep cut or open sore, be sure to treat it, keep it covered, and consider staying out of the water until it's healed. And for heaven's sake, don't think the ocean water will disinfect it! The ocean can cure your soul, but not your scrape.

Seawater Is Just Salty Water

If you've ever had a mouthful of seawater, you know it's salty. In fact, the average concentration of salt in ocean water is about 3.5 percent. However, seawater is much more than salty water. In addition to sodium chloride (salt), seawater contains magnesium, sulfur, calcium, potassium, carbon dioxide, nitrogen, oxygen, hydrogen, and other chemicals and organic compounds, along with microscopic bacteria, viruses, and fungi.

Understanding the composition of seawater is important to appreciating how fragile it is — too little or too much of any single component can throw off the delicate balance needed for the marine plants and animals that live in it — to them, water pollution is air pollution, too.

REMEMBER

Not all seawater has the same concentration of microorganisms. Far more microbes live near the surface and at the bottom of the ocean than in between. A single drop of seawater near the surface of the ocean may contain up to a million microorganisms. Think about that the next time you swallow a gulp of seawater or it shoots up your nose.

Waterspouts Are Tornadoes Over Water

This "myth" is actually half true. Some waterspouts *are* tornadoes over water. The tornado forms on land, then moves out over the water. These *tornadic waterspouts* are associated with stormy weather, lightning, hail, and lots of waves. Also, they develop from the top down — starting as swirling winds in the sky, then extending down to ground level.

Fair weather waterspouts are different. These swirling columns of mist and air form over the water in relatively calm weather, starting near the surface and growing their way up into the sky. Another distinction is that they move very little, so you can usually avoid one. The best approach is to move at a 90-degree angle to the direction in which it's moving.

WARNING

Never move closer to investigate a waterspout. Some can be just as dangerous as tornadoes.

The Great Pacific Garbage Patch Is a Solid Mat of Plastic

The Great Pacific Garbage Patch (what's so great about it?) is a concentration of debris brought together by the swirling surface waters (gyre) between Hawaii and the U.S. mainland. Sometimes, the way it's described makes people think it's a solid mat of plastic that you can walk on and would be easy to clean up — just tow it to shore and deal with it, right?

Trouble is, it's not all plastic, and it's not *that* concentrated. It's more like a dense soup of plastic and other floating garbage in all shapes and forms, as energy from the water and sun break it into smaller pieces. (Some plastic floats, some sinks, but either way, it's a problem.)

REMEMBER

Most plastics do not biodegrade. They do, however, photo degrade, which means UV radiation breaks them down into tiny pieces called *microplastics.* Many animals eat plastic at all levels of the food web thinking it's food or because it's inside an animal they're eating. Some animals have so much of it in their gut that they die with a full stomach. In short, recycling is not the answer; we need to figure out a way to stop producing and discarding so much of this deadly stuff in the first place. See Chapter 21 for more about the problem with plastic.

All Jellies Can Sting Humans

Jellies are *cnidarians,* a word that translates to "stinging creature," but not all cnidarians sting, and among those that do sting, some are harmless to humans. One of the defining characteristics of cnidarians is their *nematocysts* (stinging cells), but many species have completely lost these stinging cells over the course of evolution. The golden jellyfish in Palau have almost no nematocysts, and they don't need them to capture food; they've evolved a symbiotic relationship with algae, which gives them the ability to derive most of their food from photosynthesis. Many other types of jellyfish have stingers that can't penetrate human skin.

WARNING

Unless you're highly skilled at identifying jellyfish, we recommend steering clear of all of them. Some, such as the box jelly, are highly venomous and potentially deadly to humans. And peeing on a jellyfish sting does *not* help, but vinegar does.

The Ocean Is Blue Because It Reflects the Color of the Sky

First, the ocean isn't just blue. Depending on how the sun and water interact and how particles in the water reflect the light, it may appear gray, brown, navy blue, turquois, or even green or red during an algae bloom.

However, if you're referring to the water itself, water is clear; it has no color. What makes the ocean appear blue is that water absorbs light with the longest wavelengths (reds and yellows) first, leaving the light with shorter wavelengths (green and blue) for us to see. Of course, light doesn't penetrate beyond about 900 meters (3,000 feet) so any deeper than that, water has no color. It's black — total darkness.

Nothing Lives in the Middle of the Ocean

The open ocean is often described as a vast desert devoid of life. While the concentration and diversity of life in the middle of the ocean is certainly much less than in shallower waters near the coasts, the open ocean still has quite a lot of life floating, swimming, and crawling around. From the millions of micro-organisms, planktons, diatoms, and more that swim in the open water, to the larger fish and whales that call the open ocean home, you can find plenty of action if you know where to look.

In fact, one area in the open ocean, in the middle of the Atlantic, is actually teeming with life — the Sargasso Sea (see Chapter 5), named after the enormous mats of sargassum seaweed floating on its surface. This seaweed is home to a multitude of baby fish, octopi, sea turtles, crabs, seahorses, and more that hide and feed among it.

REMEMBER

You can find life at the very bottom of the open ocean, beyond the depths sunlight penetrates. Bacteria feeding on chemicals spewing from hydrothermal vents on the ocean floor anchor food webs that support a diversity of life-forms including tube worms, corals, mussels, snails, shrimp, and fish.

IN THIS CHAPTER

» **Reducing your carbon footprint and use of plastics**

» **Becoming a conservation-minded consumer**

» **Casting your vote to support nature**

» **Rallying friends and family members to join the cause**

» **Supporting organizations committed to preserving nature**

Chapter **24**

Ten Ways You Can Help Preserve the Ocean

E arth has experienced five mass extinction events over the past 540 million years, the last of which occurred about 65 million years ago. All of these were due to natural, cataclysmic events. Many scientists suggest that we're in the midst of a sixth mass extinction, this one driven by human activities — overpopulation, pollution, the burning of fossil fuels, deforestation, coastal "development," irresponsible fishing and hunting, and more. The good news is that this mass extinction is avoidable.

You have the power to save the planet and the ocean. It's that simple. It's not a question of *whether* you can make a difference. *Everything* you do makes a difference. The question is whether what you do has a positive or negative impact on the planet. Every human being makes that choice, usually multiple times a day, by deciding how to travel, what to buy, how to vote, what to invest in, how to interact with nature, how to influence others, and more.

We encourage you to adopt a conservation mindset that guides you to make the right choices. You can start small but start now because a healthy ocean gives us so much: more food, safer coastlines, stable climate, and so much more including trillions of dollars' worth of economic benefit. While a healthy ocean is great for all sorts of animals, it helps one animal most of all — humans. In this chapter, we present ten ways you can make a positive impact on Earth and its ocean.

Reduce Your Carbon Footprint

Your *carbon footprint* is a measure of the carbon compounds you leave behind over the course of your life. You can reduce your carbon footprint in two ways — release less carbon and sequester (remove and store) more carbon.

REMEMBER

The excess carbon we are pumping into the atmosphere that is causing climate change is wreaking havoc on our planet and on us.

Here are several easy ways to reduce your carbon footprint:

>> Wash your clothes in cold water and hang them out to dry.

>> Turn down the temperature on your water heater.

>> Eat lower on the food chain — mostly fresh vegetables and fruits, less meat, and way less processed food. In addition to reducing your carbon footprint, you'll feel a whole lot better.

>> Buy only as much food as you're going to eat. If food waste were a country, it would be the third largest emitter of greenhouse gases after the U.S. and China.

>> Drive less. Walk, ride your bike, take public transportation, or carpool.

>> If you're in the market for a new car, consider an electric car or a hybrid. Eight years ago, we joined the electric car revolution and haven't looked back. Electric cars are faster, quieter, have virtually no maintenance, are cheaper to fuel up, and are waaaaaaay more fun to drive — not to mention the fact that they reduce carbon emissions and pollutants.

>> When planning vacations, look for destinations that have green certifications or use websites that specialize in conscientious travel to plan your trip.

>> Plant a tree and avoid, as much as possible, chopping down any.

Steer Clear of Plastics

Plastics break down, but they never biodegrade, so they remain in the environment for thousands of years and usually end up in the ocean. This has huge impacts on the health of aquatic ecosystems and thus all the creatures that rely on them, including us. Here are some ways to cut down on their use:

>> Bring your own reusable bags when you head to the store.

>> Avoid single-use (disposable) plastic items, such as plastic plates, cups, utensils, and straws. These are only used for an average of two minutes, but they stick around forever!

>> Don't buy bottled water. Use a refillable bottle.

>> Use glass or stainless-steel storage containers instead of plastic or (worse) plastic bags.

>> Buy quality clothing made of natural materials that you plan to wear for a long time and opt for vintage or recycled clothing. Some clothing is even made out of plastic; avoid that.

>> Buy laundry and cleaning products in the form of concentrated dry tablets instead of liquids in plastic bottles. You can now buy tablets to mix your own cleaning products. Simply drop a tablet into a reusable spray bottle, add water, and voila . . . you have your glass, bathroom, or multi-surface cleaner. Oh, and they're cheaper too, so it's a win-win.

>> Recycle recyclable plastics.

>> Boycott products that contain microbeads, which may be in toothpaste, sunscreen, facial scrub, body wash, cosmetics, and other personal care products. Common microbead ingredients include Polyethylene (PE), Polyethylene terephthalate (PET), Nylon (PA), Polypropylene (PP), and Polymethyl methacrylate (PMMA).

>> Eat out less, especially at food joints that use a lot of plastic and Styrofoam.

Make Sustainable Seafood Choices

REMEMBER

Hundreds of millions of people rely on seafood to feed their families. Without seafood as a source of protein for the world's population, a lot more people will go hungry leading to major economic and social upheaval.

Sustainable seafood is caught or farmed in ways that minimize negative environmental and social impacts. Here are a few ways to make sustainable seafood choices:

>> Buy seafood labeled with the Ocean Wise or Marine Stewardship Council (MSC) stamp of approval.

>> Check the Monterey Bay Seafood Watch website at www.SeafoodWatch.org or install the Seafood Watch app on your smartphone for advice on selecting sustainable seafood.

>> Eat lower on the food chain by consuming smaller fish, such as sardines, anchovies, mackerel, and herring. However, check first, because even these smaller fish can be overfished.

>> Support local fisheries. For example, when you're on the East Coast, eat East Coast seafood.

>> Before ordering seafood at a restaurant, ask where they get their fish — any good seafood restaurant will know. If they don't, choose something else. And never eat bluefin tuna; they are close to extinction.

Use Ocean-Friendly Sunscreen

Sunscreens can be divided into two categories — chemical and physical. Chemical sunscreens absorb ultraviolet (UV) rays and convert them into heat that dissipates from your body. Sounds good, right? Well, your body absorbs some of those chemicals, and their impact on human health is unknown. Worse is that they wash off in the water and are toxic to some marine larvae and coral, and coral are having enough trouble as it is. More than half the world's coral reefs are disappearing with terrible consequences for our ocean and for us. Physical sunscreens also contain chemicals, but they don't penetrate the skin and are not harmful to ocean life.

REMEMBER

Read the ingredients printed on the label. Avoid sunscreens that contain avobenzone, octisalate, and chemicals with other names that leave you tripping over your tongue. Choose sunscreens that use zinc oxide or titanium oxide as the main ingredient. Products with larger particles (non-nano) of the active ingredient may also be safer.

Don't Buy Products That Exploit Marine Life

Menhaden . . . you've probably never heard of this little fish. It's a type of herring that's one of the most important fish in the ocean because it bridges the gap between zooplankton and phytoplankton and larger fish. In the Atlantic and Gulf it's vital to the food web, yet it has been mercilessly harvested for decades for its oil, which ends up in all sorts of products such as fish oil supplements, lipstick, fertilizer, and dog food. Menhaden, cod, and other fish stocks have been decimated as cheap additives to all sorts of products. But that's not all. Shark cartilage is used in joint supplements, and squalene from shark liver is used in beauty products. Steer clear of these and make sure you check the label before you buy (squalene can also come from plants and that version is A-OK).

WARNING

Fish oil supplements are the worst! Worried about Omega 3s in your diet? Well, all those products, from supplements to eggs marketed as omega 3-enhanced, are usually produced by adding fish oil to them (chickens eating fish? *gross*).

If you're in the market for an omega 3 supplement, skip the middleman and get your omega 3s where the fish get theirs — from algae. You can find omega 3 supplements made from algae instead of fish oil. Why decimate fish stocks when you can just get it from the source? Plus, you won't have fishy burps with the algae supplements!

Vote for the Ocean with an Environmental Conscience

One of the greatest powers we have in a representative democracy is the power to vote. Regardless of which political party you favor, conservation, clean air, and fresh water should unify us, not divide us because we are all in this together. Send a message to your local, state, and national politicians that protecting the ocean is a top priority for you and your family, then vote for the candidates who are most committed to saving the planet . . . and our beloved ocean.

Defend Your Drain: Use Natural Products

Everything that flows down your drain — detergents, bath soaps, shampoos, conditioners, toilet bowl cleaners, drain cleaners, and more — has the potential to end up in rivers, lakes, streams, and ultimately the ocean, posing a threat to everything living in those waters. Water is the most precious substance on Earth; water pollution is devastating not only to the health of wildlife but to human health as well. To reduce the amount of harmful products going down your drains, take the following precautions:

>> Use soaps, creams, detergents, and cleaning products that are less harmful to the environment. A good rule of thumb is to avoid products labeled with the names of any ingredients you can't pronounce. Consult resources such as the Environmental Working Group website at www.ewg.org to find environmental ratings for products.

REMEMBER

Phosphates in detergents are particularly bad, causing harmful algae blooms. They've been banned from laundry detergents but are still permitted in dishwashing detergents and other products (even ones we put directly on our bodies, such as soap and shampoo).

>> Don't flush pills or pour grease, oil, paints, solvents, or anything else down the drain that doesn't belong in there.

>> Don't flush wipes, even so-called flushable wipes, or flushable kitty litter (or cat poop — remember Chapter 15!).

Protect Your Local Watershed

Regardless of how far away you are from the ocean, your local watershed is connected to it. (A *watershed* is a land area that channels rainfall and snowmelt into creeks, streams, and rivers and ultimately to lakes, bays, and the ocean.) Everything flows downstream, and we mean *everything*, so protecting the watershed means being careful about everything you put down your drain, on your lawn and garden, and on your sidewalks and driveway; keeping your vehicle in good repair, so it doesn't leak oil, gas, or antifreeze onto pavement; conserving water; recycling yard waste and using a mulching lawn mower; and dumping pet waste in the trash.

REMEMBER

Again, the quality of our water is directly tied to the quality of our health. Ensuring that water is up to snuff is not something we can just assume but something we have to be active participants in. The health of our communities, our families, and ourselves depends on it.

Join the EarthEcho Water Challenge at www.monitorwater.org — one of the largest citizen science water quality testing programs in the world. With a simple kit, spend some quality time outdoors understanding your watershed, testing the water, and taking action to protect and restore it.

Make It a Family Affair

Growing up as the third generation of a family dedicated to the exploration and protection of our natural world, I (Philippe) have always believed in the power of young people to drive change. In fact, my grandfather, Jacques Cousteau, always told me, "Before we talk about conservation, we must talk about education." Throughout my career, I have made this my guiding principle. As the founder of EarthEcho International, a leading youth environmental educational organization, I have seen young people all over the world make tremendous change in their communities. Our youth leaders have passed laws, raised critical funds, started movements to protect land, founded successful businesses that help people and the planet, and so much more. So, one of the best things you can do for the environment is to make helping the ocean a family mission.

When you talk to your kids (or kids, when you talk to your parents) about the environment, connect the dots for them about how local bodies of water make much of our daily lives possible, from cooking food to brushing our teeth. Or how reusable shopping bags can help save their favorite marine animal. Even the smallest journey can spark exploration and discovery that's transformative.

Whether it's a small step in your household or participation in global programs, empowering kids to take part in solutions fuels a sense of pride that can inspire a child for a lifetime. Awareness does not lead to action; action leads to awareness. Tapping into your child's inherent curiosity with activities that are accessible and that deliver tangible results are excellent starting points. Action doesn't have to be complicated; it just needs to be part of the equation. That's the most effective way to tap into the inner problem solver in every child.

REMEMBER

Today, youth understand far more about the state of our planet than we adults may think. They're determined to create a different relationship with the natural world that focuses on solutions, not excuses. As parents, mentors, educators, and community leaders, we can help them on their journey, and they can help us on ours.

Join and Support Ocean Conservation Organizations

Our favorite organization is the one we started. EarthEcho International (`www.earthecho.org`) has become a leading global environmental education organization and has worked with over 2 million people in 146 countries to help them become leaders who are fighting for a healthy environment every day. Here are several others we really like:

Algalita (`https://algalita.org`)

Beneath the Waves (`https://beneaththewaves.org`)

Big Blue and You (`https://bigblueandyou.org`)

Clearwater Marine Aquarium Research Institute (`https://mission.cmaquarium.org`)

Conservation International (`www.conservation.org`)

Coral Reef Restoration Foundation (`www.coralrestoration.org`)

Environmental Working Group (a great resource about products) (`www.ewg.org`)

League of Conservation Voters (`www.lcv.org`)

Mapping Ocean Wealth (`https://oceanwealth.org`)

Mission Blue (`https://mission-blue.org`)

Monterey Bay Seafood Watch Guide (`www.seafoodwatch.org`)

Mote Marine Laboratory & Aquarium (`https://mote.org`)

Ocean Conservancy (`https://oceanconservancy.org`)

Oceans 2050 (`www.oceans2050.com`)

The Ocean Foundation (`https://oceanfdn.org`)

Ocean Heroes Network (`https://oceanheroeshq.com`)

Ocean Unite (`www.oceanunite.org`)

Only One (`www.only.one`)

The Outlaw Ocean Project (`www.theoutlawocean.com`)

Plant a Million Corals (`http://plantamillioncorals.org`)

Pristine Seas (`www.nationalgeographic.org/projects/pristine-seas`)

Schmidt Ocean Institute (`https://schmidtocean.org`)

SeaLegacy (www.sealegacy.org)

Seattle Aquarium (www.seattleaquarium.org)

Surfrider Foundation (www.surfrider.org)

Sustainable Ocean Alliance (www.soalliance.org)

Underwater Earth (www.underwater.earth)

Waterkeeper Alliance (https://waterkeeper.org)

World Oceans Day (https://worldoceansday.org)

World Wildlife Fund (www.worldwildlife.org)

Woods Hole Oceanographic Institution (www.whoi.edu)

Youth Ocean Conservation Summit (www.yocs.org)

Index

About the Authors

Ashlan and Philippe Cousteau are environmental advocates and filmmakers with a deep mutual passion for exploration and storytelling. Their mission is to inspire, entertain, and motivate people to solve the critical issues facing humans, wildlife, and the planet. The adventurous couple collaborate frequently; from co-narrating the virtual reality experience *Drop in the Ocean,* to co-starring in the award-winning Travel Channel series *Caribbean Pirate Treasure,* hosting *Nuclear Sharks* for Discovery's Shark Week and more, Ashlan and Philippe believe that making our world a better place can be a life-changing adventure.

In addition to their work together, Philippe and Ashlan are respected journalists and explorers in their own right. Ashlan was a reporter and fill-in anchor for *E! News* and special correspondent for *Entertainment Tonight* for over a decade. But her endeavors go beyond the small screen. Ashlan served as host for the UN's convention on migratory species in Quito, Ecuador. She anchored former Vice-President Al Gore's internationally live broadcast of *Climate Reality,* has presented at the Society of Environmental Journalists, SXSW Eco, and was a speaker at TEDx Scott Base (the first TED conference to take place in Antarctica), and many more international events. She has been recognized on numerous talk shows and in media for her work in conservation and has explored all seven continents.

Philippe is the third generation of the legendary Cousteau family. He is a multi-Emmy-nominated TV host and producer as well as an author, speaker, and social entrepreneur who has established himself as a prominent leader in the environmental movement. He is the host and executive producer of *Awesome Planet,* a weekly syndicated television series, and served as a special correspondent for CNN for several years where he hosted award-winning shows including *Going Green* and *Expedition Sumatra.* He is also the author of several award-winning children's books, including *Follow the Moon Home* (Chronicle Books), *Going Blue, Make a Splash* (both by Free Spirit Publishing), and *The Endangereds* (HarperCollins Publishers). His conservation efforts are focused on solving global social and environmental problems. In 2004, he founded EarthEcho International — a leading environmental education organization which is building a global youth movement for the ocean and has reached two million youth in 146 countries.

Ashlan and Philippe are the global ambassadors for Aqua Lung scuba diving, and they serve on the National Council of the World Wildlife Fund, the Environmental Media Association, and the Ocean Unite Network. When not traveling or filming around the world, they live in Los Angeles, CA, with their daughter, Vivienne; rescue dog, Kenai Cousteau; and two chickens, Heidi Plume and Cindy Cluckford.

Dedication

From Ashlan: To all the kids out there (young and old) who are captivated by the sea. To my husband, who gave me the confidence to write a book about the ocean. And to our daughter, Vivienne, who has made us fall in love with nature all over again through her innocence, kindness, and curiosity.

From Philippe: To my darling Ashlan, you wrote most of this book, and I am so proud of the passion, humor, and integrity you brought to each and every page. Each day with you is an extraordinary adventure, and I never cease to be inspired by your dedication to making this world a better place. Thank you for letting me share this journey with you.

Authors' Acknowledgments

It takes a village to publish a book, and this book is no exception. Thanks to Wiley executive editor Lindsay Lefevere for guiding this project, and to our co-writer, Joe Kraynak, and our researcher, Hayley Charlton-Howard, who were instrumental in gathering content and collaborating on the writing and editing. Thanks also to Chrissy Guthrie, Vicki Adang, and Michelle Hacker for polishing our prose and carefully shepherding our text, illustrations, and photos through the production process. Completing this project during a pandemic with multiple ongoing projects and a very active toddler at home would have been almost impossible without the skilled assistance of all these fantastic professionals.

Thank you to our wonderful and talented friends who lent their incredible photographs to us to make this book come alive: Cristina Mittermeier, Paul Nicklen, Michael Muller, John and Dan Cesere, Keith Ellenbogen, Anamaria Chediak, John Weller, Romona Robbins, Shane Reynolds, Sheree Marris, Jamie Seymour, Laurent Ballesta, Hal Wells, Jamal Galves, Loren McClenachan, Jacques Renoir, Jason Hall-Spencer, Martin Attrill, Carlie Wiener, Carlos Duarte and Pier Nirandara. And to the unwavering work of Woods Hole Oceanographic Institution, Mote Marine Laboratory & Aquarium, Schmidt Ocean Institute, WILDCOAST, the Seattle Aquarium, Underwater Earth, University of Plymouth, Institute of Marine Research, and of course the National Oceanic and Atmospheric Administration (NOAA).

Thank you to our technical editors, Marc J. Alperin and Stephen R. Fegley from the University of North Carolina at Chapel Hill's Department of Marine Sciences, for contributing their deep knowledge of all things ocean to fact-checking our manuscript and offering their expert insight. Go Heels! (Ashlan is a proud UNC-CH alumna 2002).

And special thanks to the teachers, ocean advocates, conservationists, youth leaders, and all the people who dedicate their lives to science and expanding the understanding of our natural world. This book is also dedicated to you: Together we can build a hopeful future by protecting and restoring our mighty ocean.

The book is not affiliated with or related to The Cousteau Society or its founder Jacques-Yves Cousteau or of his work.

Publisher's Acknowledgments

Executive Editor: Lindsay Sandman Lefevere

Managing Editors: Vicki Adang, Michelle Hacker

Editorial Project Manager and Development Editor: Christina N. Guthrie

Technical Editors: Marc J. Alperin, Department of Marine Sciences, University of North Carolina at Chapel Hill, and Stephen R. Fegley, Institute of Marine Sciences, University of North Carolina at Chapel Hill

Proofreader: Debbye Butler

Production Editor: Tamilmani Varadharaj

Cover Image: © Keith Ellenbogen

Author Photo: © Voyacy

Full Image Credits for Public Domain Images

Figure 5-11: NOAA Photo Library, Source: Sam Farkas, NOAA OAR Photo Contest 2014. `https://www.flickr.com/photos/noaaphotolib/19590492028`. Licensed under CC BY 2.0

Figure 5-16: NOAA Photo Library, Source University of Washington; NOAA/ OAR/OER. `https://www.flickr.com/photos/noaaphotolib/5277263409/in/album-72157635360690997/Deep-sea` coral reefs. Licensed under CC BY 2.0

Figure 7-1: U.S. Department of Energy, Pacific Northwest National Laboratory `https://www.pnnl.gov/science/images/highlights/biology/Lipton_bacterioplankton.jpg`. Public Domain

Figure 7-2: Image courtesy of NOAA Bioluminescence and Vision on the Deep Seafloor 2015 `-https://oceanexplorer.noaa.gov/edu/themes/bioluminescence/multimedia.html#cbpi=media/multimedia-chrysogorgia.html`. Public Domain

Figure 7-6: MEB back, Source: Zatelmar. `https://commons.wikimedia.org/wiki/File:MEB_back.png`. Licensed under CC BY 3.0.

Figure 7-7: NOAA Photo Library, Source: Matt Wilson/Jay Clark, NOAA NMFS AFSC. `https://www.flickr.com/photos/noaaphotolib/9861767625/in/album-72157653896047593/`. Licensed under CC BY 2.0

Figure 7-8: NOAA Photo Library, Source: Matt Wilson/Jay Clark, NOAA NMFS AFSC. `https://www.flickr.com/photos/noaaphotolib/9861752254/in/album-72157653896047593/`. Licensed under CC BY 2.0

Figure 8-2: NOAA's America's Coastlines Collection Captain, Source: Albert E. Theberge, NOAA Corps (ret.) `http://www.photolib.noaa.gov/htmls/line2709.htm`. Licensed under CC BY 4.0

Figure 8-3: NOAA Photo Library, Source: Dr. John R. Dolan, Laboratoire d'Oceanographique de Villefranche; Observatoire Oceanologique de Villefrance-sur-Mer `https://www.flickr.com/photos/noaaphotolib/16061417856/in/photolist-vbUebX-vR9ViQ-w9bRFr-qvpsm6-qthXXw-fKfFur-w87Yo9-8RMEhz`. Licensed under CC BY 2.0

Figure 9-3: NOAA Photo Library, Source: Chris Coccaro, Bonaire 2008 Exploring Coral Reef Sustainability with New Technologies.; NOAAOAROER. `https://commons.wikimedia.org/wiki/File:Reef1995_-_Flickr_-_NOAA_Photo_Library.jpg`. Licensed CC BY 4.0

Figure 9-4: NOAA Image courtesy of the NOAA Office of Ocean Exploration and Research, Deep-Sea Symphony Exploring the Musicians Seamounts. `https://`

`oceanexplorer.noaa.gov/okeanos/explorations/ex1708/dailyupdates/media/sept21-1.html`. Public Domain

Figure 9-8: Source: NOAANOSNMSFGBNMS; National Marine Sanctuaries Media Library Licensed under CC BY 4.0

Figure 9-9: NOAA Photo Library, Source: NOAANOSNMSFGBNMS; National Marine Sanctuaries Media Library.`https://commons.wikimedia.org/wiki/File:Sanc0448_-_Flickr_-_NOAA_Photo_Library.jpg`. Licensed under CC BY 4.0

Figure 9-10a: NOAA Photo Library, Source: Dr. Dwayne Meadows, NOAA_NMFS_OPR. `https://www.flickr.com/photos/noaaphotolib/5018044379/in/album-72157635438729212/`. Licensed under CC BY 2.0

Figure 9-10b: NOAA Photo Library, Source: NOAA's America's Coastlines Collection, Photograph by Julie Brownlee. `https://www.flickr.com/photos/noaaphotolib/9787410974/in/photolist-fUS3so-JNqoGK-fUT3i5-23tHBJJ`. Licensed under CC BY 2.0

Figure 9-11b: NOAA Photo Library, Source: NOAA Okeanos Explorer Program, INDEX-SATAL 2010. `https://www.flickr.com/photos/noaaphotolib/9720863894/in/photolist-fNZYaC-fNZVAC-fNHnv2-fNHoqi-fNHq7P-fNRiP1-fNZUFj-fNH4mt-K6G256-fNZCay-fNZAN9-it6p7D-22a6AGp-fNH3Yi-fNZWQs-DSv2iz-vbXCd2-JXgbQT-J8MGJG-fNHpaZ-DSv3dk-w6s7UL-JqbE7U-vbYJT4-w8R1G6-fNHqAk-Jz9ZmZ-JzbU8P`. Licensed under CC BY 2.0

Figure 10-2: NOAA Photo Library, Source: Lt. John Crofts, NOAA Corps. `https://www.flickr.com/photos/noaaphotolib/5018037479/in/album-72157635589790785/`. Licensed under CC BY 2.0

Figure 10-3: NOAA Photo Library, Source: G. P. Schmahl, NOAA FGBNMS Manager. `https://www.flickr.com/photos/noaaphotolib/5277563747/in/photolist-8UsqZm-93mTQp`. Licensed under CC BY 2.0

Figure 10-5: NOAA Photo Library, Source: NOAA's Fisheries Collection, Mandy Lindeberg, NOAA/NMFS/AKFSC. `https://www.flickr.com/photos/noaaphotolib/9787178464/in/album-72157635589790785/`. Licensed under CC BY 2.0

Figure 10-6a: Nembrotha kubaryana, Source: Nick Hobgood. `https://commons.wikimedia.org/wiki/File:Nembrotha_kubaryana_(Nudibranch).jpg`. Licensed under CC BY-SA 3.0

Figure 10-7: NOAA Photo Library, Source: Kevin Lino NOAA/NMFS/PIFSC/ESD. `https://photolib.noaa.gov/Collections/Coral-Kingdom/Pacific-Reefs/Invertebrates/emodule/759/eitem/34916`. Licensed under CC BY 2.0

Figure 11.-1: Female adult of the water flea Daphnia magna by Hajime Watanabe, Source: Hajime Watanabe, `https://search.creativecommons.org/photos/e390c066-7c25-4544-91e9-ac7b3b945f74`. Licensed under CC BY 2.0.

Figure 11-2: Nebalia bipes, Source: Hans Hillewaert. `https://commons.wikimedia.org/wiki/File:Nebalia_bipes.jpg`. Licensed under CC BY-SA 4.0.

Figure 11-6: NOAA Photo Libraray, Source: NOAA. `https://nmsflowergarden.blob.core.windows.net/flowergarden-prod/media/archive/image_library/inverts/rednightshrimpgps.jpg`. Public Domain

Figure 11-8: Very large hermit crab making queen triton shell home., Source: NOAA. `https://photolib.noaa.gov/Collections/Sanctuaries/Papa/emodule/839/eitem/36784`. Public Domain

Figure 12-1: Source: U.S. Fish and Wildlife. `https://nhpbs.org/wild/Agnatha.asp`. Public Domain

Figure 12-10: Source: NOAA Office of Ocean Exploration and Research, Gulf of Mexico 2018. `https://oceanexplorer.noaa.gov/image-gallery/welcome.html#cbpi=/okeanos/explorations/ex1803/dailyupdates/media/apr16-1.html`. Public Domain

Figure 12-17: JawFishEggs, Source: Tracy Candish `https://www.flickr.com/photos/80595542@N07/33874845733`. Public Domain 1.0

Figure 13-1: Source: NOAA. `https://www.fisheries.noaa.gov/feature-story/10-tremendous-turtle-facts`. Public Domain

Figure 13-10: Saltwater Crocodile on a river bank, Source: Paul Thomsen (WILDFOTO.COM.AU). `https://commons.wikimedia.org/wiki/File:Saltwater_Crocodile_on_a_river_bank.jpg`. Public Domain

Figure 14-9: Double-Crested Cormorant, Source: Colin Durfee `https://www.flickr.com/photos/146003125@N02/49550621046`. Licensed under CC BY 2.0

Figure 15-1: Source: Dr. Elliott Hazen NMFS/SWFSC/ERD. `https://photolib.noaa.gov/Collections/Sanctuaries/Stellwagen-Bank/emodule/836/eitem/36924`. Public Domain